Biomechanics of the Musculoskeletal System

FOCUS SERIES

Series Editor Marie-Christine Ho Ba Tho

Biomechanics of the Musculoskeletal System

Modeling of Data Uncertainty and Knowledge

Tien Tuan Dao
Marie-Christine Ho Ba Tho

iSTE

WILEY

First published 2014 in Great Britain and the United States by ISTE Ltd and John Wiley & Sons, Inc.

ISTE Ltd
27-37 St George's Road
London SW19 4EU
UK

www.iste.co.uk

John Wiley & Sons, Inc.
111 River Street
Hoboken, NJ 07030
USA

www.wiley.com

Library of Congress Control Number: 2014934406

British Library Cataloguing-in-Publication Data
A CIP record for this book is available from the British Library
ISSN 2051-2481 (Print)
ISSN 2051-249X (Online)
ISBN 978-1-84821-602-0

MIX
Paper from responsible sources
FSC www.fsc.org FSC® C013604

Printed and bound in Great Britain by CPI Group (UK) Ltd., Croydon, Surrey CR0 4YY

Contents

Preface

Biomechanics of the musculoskeletal system covers a large range of research topics using experimental and numerical approaches. *In silico* numerical models have usually been developed to describe the mechanical behavior of the musculoskeletal system under internal and external loadings. Such models allow us to better understand the mechanical behavior of the different components of the musculoskeletal system (joints, organs, tissue, etc.) and their interaction. Moreover, knowledge obtained from *in silico* model analysis and simulation could be used to help clinicians and/or engineers in their decision-making process for diagnosis, treatments, follow-ups as well as technology development for health care and bioengineering.

However, biomechanical data, used as input data of *in silico* models, are subject to uncertainties due to subject variability, technical protocol assessing experimental data and subsequently numerical processing methods. As a result, this book provides comprehensive and clear contents of the modeling of data uncertainty and knowledge of the biomechanics of the musculoskeletal system. This book is especially aimed at engineers and medical students interested in the biomedical field.

This book is divided into five chapters. Chapter 1 provides an overview of *in silico* rigid multi-bodies musculoskeletal model. Chapter 2 introduces one of the main topics of this book, the modeling of data uncertainty. Chapter 3 focuses on the knowledge modeling of the musculoskeletal system. Chapter 4 addresses some clinical applications of biomechanical and knowledge-based models for orthopedic disorders. Chapter 5 presents some practical software and tools for knowledge modeling and reasoning purposes.

Tien Tuan DAO
Marie-Christine HO BA THO
February 2014

1

Biomechanics of the Musculoskeletal System

The musculoskeletal system plays an essential role in the equilibrium and motion of the human body. Biomechanics of the musculoskeletal system uses physical laws and engineering methods to describe the mechanical behavior of the musculoskeletal system during motion. In this chapter, first, the introduction of biomechanics and related applications is presented. Second, the state of the art of knowledge in biomechanics of the musculoskeletal system, in particular the development of *in silico* rigid multi-body musculoskeletal models and their perspectives, is addressed.

1.1. Biomechanics and its applications

1.1.1. *Introduction*

Biomechanics is a research field which aims to solve biomedical or biological problems by using mechanical engineering methods, techniques and theories [HAT 74, WIN 11]. Living systems such as human musculoskeletal system or cardiovascular system are the main objects of biomechanics research study. Engineering methods range from experimental to numerical approaches. Experimental studies [KEY 65, SHA 01] aim to observe qualitatively and quantitatively the changes of biological tissues (e.g. bone, muscle, cartilage and ligament) or structures (e.g. knee) under normal and abnormal conditions. Experimental studies could be performed *in vivo* and *ex vivo* or *in vitro* conditions. *In vivo* experimentation relates to the study of whole living subject in natural environment. *Ex vivo* or *in vitro* experimentations deal with the testing of tissues isolated outside its biological surroundings of the living organism. Such experimentations are commonly performed in a culture environment. It is important to note that the characteristics and behaviors of a biological tissue/structure *in vivo* condition are completely different from those of the same tissue/structure *in vitro* or *ex vivo* conditions. Moreover, *in silico* numerical studies [REI 02, KIT 02, VEN 06] aim to model and simulate living systems to provide unobservable information of the tissue or structure under investigation such as bone stress under body loading or muscle force during motion. Moreover, numerical studies could be used to test the impact of a clinical treatment procedure (e.g. surgery or functional rehabilitation) or the impact of an implanted device (e.g. prosthesis or orthotic) on the living tissues or structures.

A biomechanics study is commonly performed in response to a basic research question or to depict its potential application for a specific case (e.g. clinical case and industrial case) as illustrated in Figure 1.1. An example of a basic research question could be how to determine the pathophysiological processes of musculoskeletal disorders. Such a basic research question allows us to better understand the functional behavior of tissues and structure. An example of an applied research study could be the application of the finite element method to predict the femoral bone stress when a femoral prosthesis is implanted to optimize the design and fabrication of the investigated prosthesis. In fact, such basic or applied research problems could be solved by using mechanical engineering methods, techniques and theories. Moreover, a biomechanics study relates to single-scale object of study (i.e. cell and molecule, tissue and organ, system, or individual or population) or multi-scale object of study.

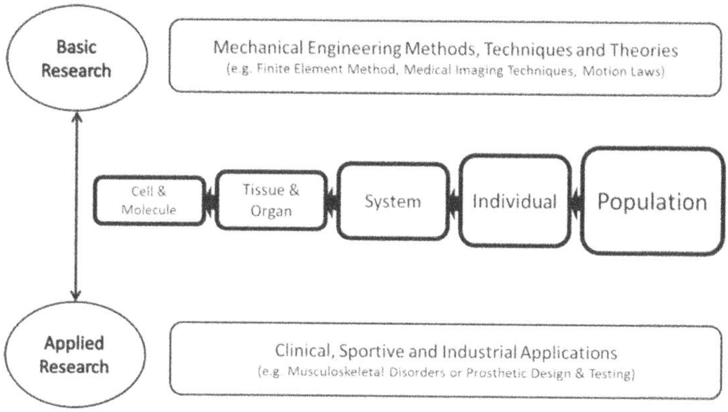

Figure 1.1. *Overview of biomechanics field of study*

1.1.2. *Applications in biomechanics*

Biomechanics studies could lead to clinical, sportive and industrial applications. A non-exhaustive list of potential applications is provided below:

– *Virtual muscle-tendon surgeries*: computer-aided modeling using *in silico* rigid multi-body dynamics could allow optimal treatment procedures to be simulated, analyzed and assessed [DEL 97]. An example of such an application is the simulation of the effect of tendon transfer on the joint behavior [RIE 97] or the muscle behavior [ASA 02].

– *Optimal design of biomedical materials and devices*: computer modeling using medical imaging and finite element method could be applied to perform the optimal design of orthopedic, dental and cardiovascular biomaterials [SLO 98]. Moreover, the effect of implanted devices (e.g. braces and prostheses) could also be assessed [PER 02]. Bioartificial devices (e.g. a liver device and a kidney device) could be designed and developed [CAR 09].

– *Assessment of gait abnormalities*: musculoskeletal disorders, such as children with cerebral palsy, have abnormal locomotion functions (e.g. stiff knee flexion). Musculoskeletal models have become customized tools to assess these abnormal functions both qualitatively and quantitatively, leading to the proposal of optimal treatment planning [ARN 01, ARN 04, ARN 05].

– *Computer-aided orbital and maxillofacial surgery*: the outcomes of facial surgery could be predicted using a patient-specific finite element model [LUB 05]. Another example is the simulation of the consequence of a surgical procedure [BUC 07].

– *Detection and prediction of preterm deliveries*: uterine electromyography (EMG) and the data mining method could be used to detect and predict the preterm deliveries, leading to a reduction in the risk of death and disabilities/impairment for premature babies [DIA 09, HAS 10].

– *Performance sportive analysis*: using different biomechanics techniques (e.g. three-dimensional (3D) motion capture, force plates, and surface electromyography), qualitative and quantitative assessments of sportive activities or exercises could be performed in order to improve the performance or prevent the risk of injury for non-professional and professional athletes [CHA 97, SPE 05, BUR 06].

– *Electrical energy harvesting*: a walking model was developed to control a wearable, knee-mounted energy harvester device to produce electrical energy with minimal user effort [KUO 05, DON 08].

– *Early diagnosis of degenerated intervertebral discs (IVD)*: lower back pain is one of the most chronic musculoskeletal disorders. Degenerated IVD is one of the possible causes of this disease. Its early diagnosis could make it possible for a better clinical outcome. Advanced medical imaging (e.g. T2 mapping and diffusion-based magnetic resonance imaging (MRI)) and image processing techniques could be used to analyze and depict the IVD changes at the tissue level, leading to early detection of the degeneration state [HAU 04, DAO 13].

1.2. Biomechanics of the musculoskeletal system: current knowledge

1.2.1. *Introduction*

Biomechanics of the musculoskeletal system is a specific branch of biomechanics, which focuses on the studies of the behavior of isolated tissues and structures (e.g. bones and segments, muscles and tendons, ligaments, cartilage, nerves and joints) as well as on the interaction between these tissues to create stability and motion functions. The objective of such a study is to provide substantial insights into the physiological and pathophysiological processes of the musculoskeletal system in the normal and pathological cases, respectively.

This section aims to describe the current knowledge extracted from basic or applied research studies on the interaction of tissues using mechanical engineering methods, techniques and theories.

Musculoskeletal models are commonly used to study the interaction of tissues. From a mechanical engineering point of view, there are two approaches for developing a musculoskeletal model as illustrated in Figure 1.2. The first approach relates to the rigid multi-body dynamics using tissue properties and Newton's laws of motion to describe the kinematic and dynamic behavior of the musculoskeletal system. The second approach deals with deformable modeling using tissue properties and finite element methods to study the structure interaction with and without fluid consideration under normal and abnormal loading conditions. In this chapter, we focus only on the rigid multi-body modeling. Current knowledge of this modeling approach is addressed in the following section.

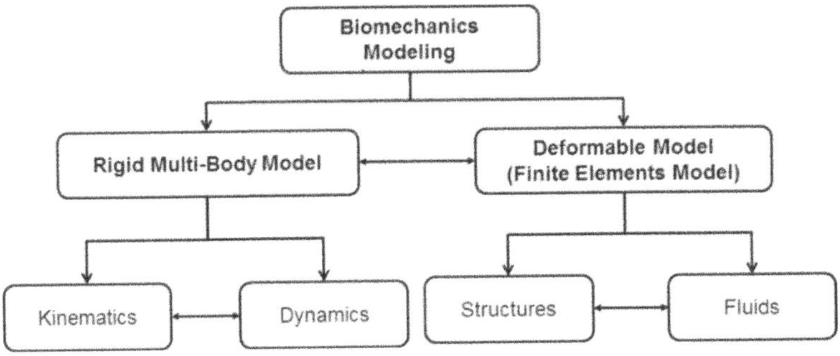

Figure 1.2. *Overview of musculoskeletal models and their interaction*

1.2.2. *Rigid multi-body musculoskeletal modeling*

In the framework of rigid multi-body dynamics, a 3D musculoskeletal model could be a generic parameterized model or a patient-specific model. The generic parameterized model uses an available model provided by musculoskeletal modeling software to scale and calibrate all properties for a specific subject. This approach reduces significantly the development time and effort. The patient-specific model uses common medical images to create individualized geometries and properties of the subject/patient under investigation, leading to more accurate simulation results. In fact, the development of a 3D musculoskeletal model requires advanced modeling knowledge and skills. Moreover, this development process is very time-consuming. For these reasons, the use of musculoskeletal modeling software is an efficient solution, especially in the case of clinical application where the decision-making needs to be performed quickly and with minimum effort. The next section addresses commonly used rigid multi-body musculoskeletal modeling software in the scientific community.

1.2.2.1. *Modeling software*

There are many pieces of modeling pieces of software, which could be used to develop generic parameterized or patient-specific musculoskeletal models. The main characteristics of commercial musculoskeletal modeling software are given in Table 1.1. There are three pieces of commercial software (AnyBody, LifeMod and software for interactive musculoskeletal modeling (SIMM)). All these pieces of software allow 3D musculoskeletal models to be developed and analyzed. The setup process of each model could be done through specific scripting language (AnyScript) or graphical user interfaces. Kinematics, kinetics and muscle forces could be computed and analyzed. Only SIMM software allows the real-time simulation to be performed using a motion analysis system. User-specific routines could be developed using automatic dynamic analysis of mechanical systems (ADAMS) script for LifeMod software.

Characteristics	AnyBody[1]	LifeMod[2]	SIMM[3]
Type	Commercial	Commercial	Commercial
Society	AnyBody Technology (Denmark)	BRG (USA)	MusculoGraphics (USA)
Analysis	3D	3D	3D
Model setup	AnyScript	Graphical user interface	Graphical user interface
Kinematics	Inverse kinematics (skin-based markers)	Inverse kinematics (skin-based markers)	Inverse kinematics (skin-based markers, joint angles)
Kinetics	Inverse dynamics	Inverse dynamics	Inverse dynamics
Muscle model	Hill-based	Closer loop Hill-based	Hill-based
Muscle forces	Static optimization	Static optimization	Static optimization
Real time			Motion Analysis
Individualized model		Bone geometries (CT, MRI)	Bone geometries (CT, MRI)
User routine		ADAMS script	

Table 1.1. *Commercial musculoskeletal modeling software*

1 http://www.anybodytech.com/.
2 http://www.lifemodeler.com/ – End of distribution from 2012.
3 http://www.musculographics.com/.

The main characteristics of open-source musculoskeletal modeling software are given in Table 1.2. These software tools provide two-dimensional (2D) and 3D analysis of the musculoskeletal system. One of the most widely used pieces of open-source software is the OpenSIM. This piece of software provides all programmable libraries and a graphical user interface to develop, simulate and analyze musculoskeletal models during motion. User-specific extensible routines could be developed using C++ code. In particular, OpenSIM allows subject- or patient-specific models to be developed in a flexible development environment. In the following section, the creation and analysis workflow of a generic parameterized musculoskeletal model is presented.

Characteristics	BodyMech[4]	OpenSIM[5]
Type	Open Source	Open Source
Developer	Jaap Harlaar (VU University, Netherlands)	Scott Delp (U. Stanford, USA)
Analysis	2D	3D
Model setup	Matlab Script	C++ Code XML script
Kinematics	Inverse kinematics (skin-based markers)	Inverse kinematics (skin-based markers, joint angles)
Kinetics	Inverse dynamics	Inverse dynamics
Muscle model		Hill-based
Muscle forces		Static optimization Computed muscle control
Individualized model		Bone geometries (CT, MRI)
User Routine	Matlab script	C++ script

Table 1.2. *Open-source musculoskeletal modeling software*

4 http://www.bodymech.nl/.

5 https://simtk.org/home/opensim.

1.2.2.2. *Modeling hypotheses*

The modeling of the biological tissues and systems plays an important role in the future personalized medicine. Using *in silico* rigid multi-body musculoskeletal models, clinical treatment outcome effects could be simulated and analyzed. However, with the current knowledge and technologies, the musculoskeletal model is only a simplified representation of the real human biological system (as illustrated in Figure 1.3). The biofidelity of a numerical model depends on the modeling assumptions.

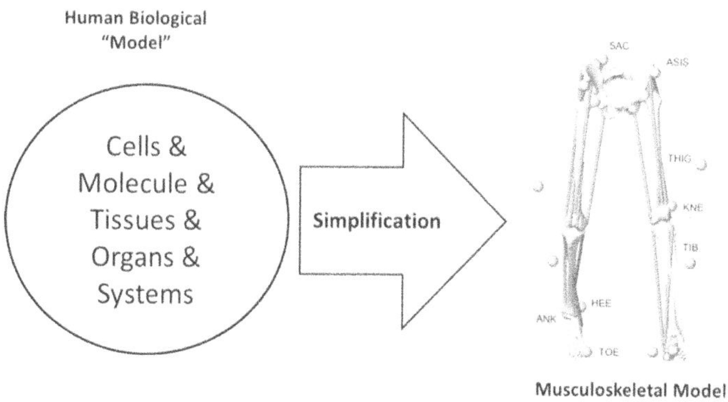

Figure 1.3. *Simplification of a musculoskeletal model*

To mathematically model and numerically simulate the biological system, modeling assumptions need to be made. Some modeling assumptions are presented below:

– *Hypothesis 1*: the central and peripheral nervous systems are neglected. Thus, the neural control mechanism is replaced by the optimization principle assuming that the human motion behavior is controlled to minimize some physiological criteria such as energy or muscle force and activation.

– *Hypothesis 2*: each body segment is a rigid body without deformation under internal and external loadings. This assumption could be neglected in the case of a deformable musculoskeletal model.

– *Hypothesis 3*: segmental mass concentrates on the barycenter of the segment. In real conditions, the position of center of mass is moved dynamically and based on the motion.

– *Hypothesis 4*: the motion of soft tissues (e.g. skin, fat and muscles) is neglected.

– *Hypothesis 5*: the sliding translation of one bone on the other bone (e.g. the case of knee joint) is neglected.

– *Hypothesis 6*: there is no muscle coordination between antagonist and agonist muscles.

1.2.2.3. *Creation and analyze workflow of a generic multi-rigid musculoskeletal model*

The use of a generic parameterized musculoskeletal model consists of the following steps: (1) model selection, (2) geometrical and anthropometrical scaling, (3) rigid multi-body dynamics and (4) partial validation. This workflow is illustrated in Figure 1.4.

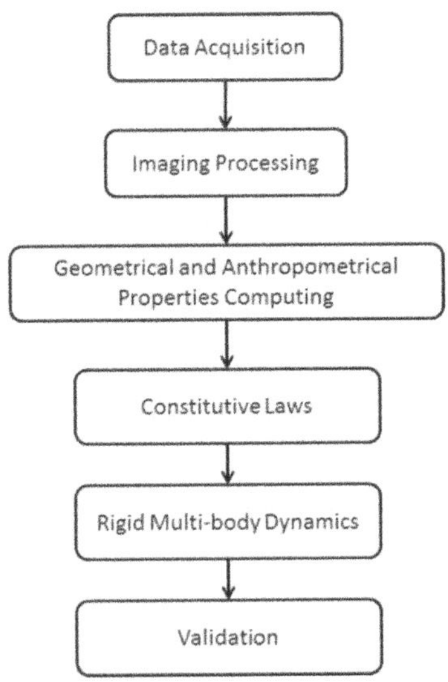

Figure 1.4. *Creation and analyzed workflow of a generic musculoskeletal model*

a) *Model selection*

This step relates to the selection of an available musculoskeletal model from the model database of the modeling software used. For example, SIMM software provides a full-body model. OpenSIM software provides different models (e.g. a lower limb model, upper limb model and lumbar spine model). Examples of OpenSIM models are illustrated in Figure 1.5.

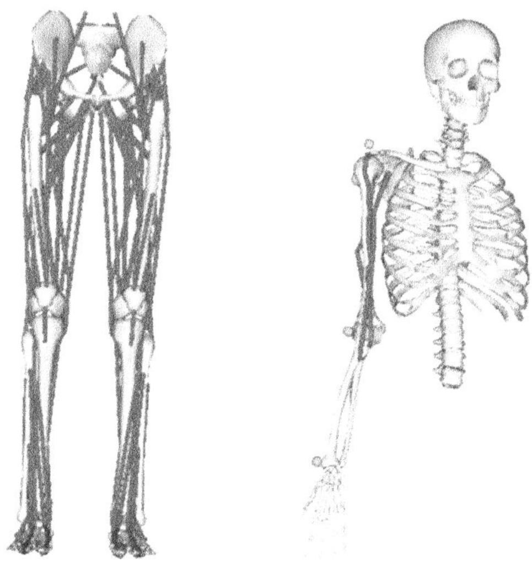

Figure 1.5. *Example of OpenSIM lower limb and arm models*

b) *Geometrical and anthropometrical scaling*

This step aims to scale the selected generic model using subject- or patient-specific data. Anthropometrical properties (e.g. subject height, body mass and body segment inertial parameters (BSIPs)) could be measured. Specific bone deformation (e.g. femoral anteversion angle) could be performed using SIMM software. Skin-based marker position could be used to scale the shape of the model using OpenSIM software.

c) *Rigid multi-body dynamics*

This step aims to apply rigid multi-body dynamics algorithms to perform simulation. LifeMod software uses inverse dynamics, static optimization, and forward dynamics to compute joint kinematics, joint kinetics and muscle forces. SIMM, AnyBody and OpenSIM only use inverse dynamics and static optimization. In particular, OpenSIM uses some specific algorithms such as the residual reduction algorithm and computed muscle control to minimize the model error and to estimate age-related muscle forces. Inverse dynamics need external motion data to compute the joint moment. We could use available motion data provided by the software used. Moreover, subject- or patient-specific experimental motion data could be obtained using 3D motion capture systems such as video converter (VICON) or motion analysis systems. Then these data could be imported into the software to perform the simulation. Kinetics data such as ground reaction forces could be used as boundary conditions of the model. In this case, available data or measured data using the force plate system could be used. For the muscle force estimation, the Hill-type muscle model is commonly used in most of the modeling software.

d) *Partial validation*

This step aims to verify the model accuracy and the simulation results. First, the simulation results have to be checked to pick out abnormal behaviors (e.g. large or unexpected joint moment or muscle forces). Second, simulation results could be compared to experimental data to validate the model. For example, experimental joint force could be used in the case of an implanted joint device to validate the joint loading patterns both qualitatively and quantitatively. EMG signals could be used to validate qualitatively the estimated muscle force patterns. However, it is important to note that current models have been partially validated. The full validation is still challenging.

Further investigations need to be performed to provide more experimental data for such validation purposes [DAO 09].

From a research point of view, the use of a generic parameterized musculoskeletal model is not sufficient due to the constraints and limitations of the modeling software used. Researchers always prefer to develop their own model to answer a specific research question. Moreover, for a clinical application, the musculoskeletal model needs to be subject- or patient-specific. Thus, the development of a rigid multi-body musculoskeletal model is an important task for biomechanical engineers or researchers. In the next section, the development workflow of a subject- or patient-specific rigid multi-body musculoskeletal model is presented. The current knowledge of each modeling step is also presented with examples.

1.2.2.4. *Development workflow of a subject-specific rigid multi-body musculoskeletal model*

The development of a subject- or patient-specific rigid multi-body musculoskeletal model consists of the following steps: (1) data acquisition, (2) imaging processing, (3) computation of geometrical and anthropometrical properties, (4) definition of tissue constitutive laws, (5) rigid multi-body dynamics and (6) partial validation. The development workflow of a subject-specific model is illustrated in Figure 1.6. The difference between the creation workflow and the development workflow of a generic model is shown in Figure 1.6.

a) *Data acquisition*

Medical imaging techniques such as MRI and computed tomography (CT) have become standard tools to obtain subject- or patient-specific musculoskeletal geometries. Data acquisition is generally performed in a medical center under the supervision of medical technicians and radiology experts. Specific imaging protocols need to be developed for each tissue (e.g. bone, muscle and cartilage) to maximize the tissue contrast and reduce the signal-to-noise ratio.

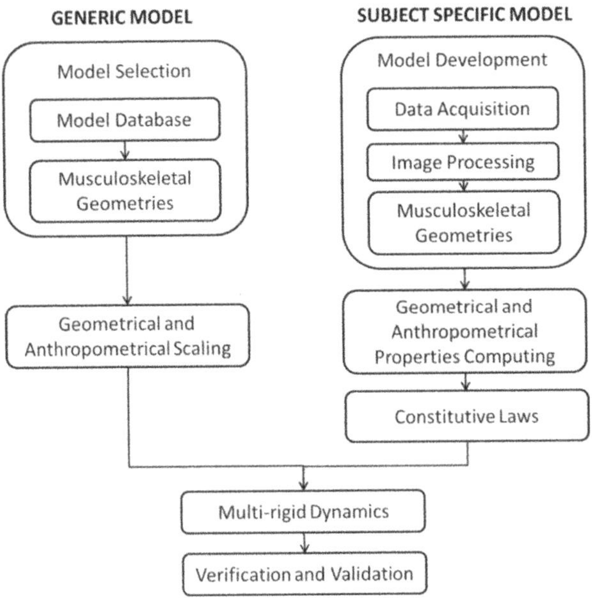

Figure 1.6. *Creation and development workflows of generic- or subject-specific musculoskeletal models*

CT is based on X-rays to produce tomographic anatomical images of an area of interest of the body. The objective of a CT protocol is to minimize the effective radiation dose with the best image quality. The CT technique is a good choice for hard tissue such as bone. The acquisition time of a CT acquisition is very fast. With current technologies, the acquisition time of CT acquisition for lower limb-structures is around 30 s. An example of CT images of lower-limb structures is shown in Figure 1.7.

MRI uses the magnetization principle of atomic nuclei to obtain anatomical images of the tissue of interest. MRI images provide good contrast for the soft tissue (e.g. muscle, skin and fat). The current MRI magnetic field strength used in clinics is 1.5 Tesla or 3 Tesla. The acquisition time of an MRI scan depends on each specific case. An example of an MRI image of the lumbar spine is shown in Figure 1.8.

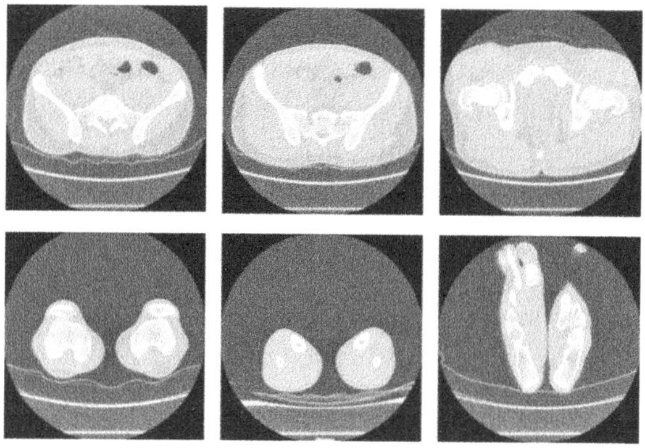

Figure 1.7. *CT images of the lower-limb structures*

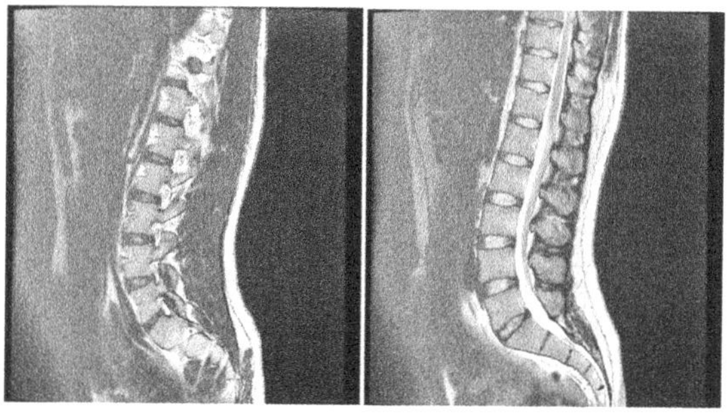

Figure 1.8. *MRI images of the lumbar spine*

b) *Imaging processing*

Based on the 2D anatomical images of each tissue of interest, imaging processing techniques are commonly applied to (1) segment the tissue of interest from surrounding tissues and (2) reconstruct the tissue of interest in 3D.

Image segmentation aims to assign a label to a group of pixels representing a tissue of interest from the raw images. Segmentation could be done using manual, semi-automatic or automatic methods. The threshold-based automatic method is commonly used for the segmentation of bone tissue with CT images. An example of automatic segmentation using the threshold principle and CT images is shown in Figure 1.9. Manual or semi-automatic methods are used for the segmentation of soft tissues with MRI images. There are many commercial and open-source software tools available to perform the image segmentation. A summary of the main characteristics of some commercial and "home-made" tools is given in Table 1.3. All tools could read digital imaging and communications in medicine (DICOM) image format. For open-source tool, we could use 3D Slicer[6] to perform the segmentation.

Based on the 2D segmented images, reconstruction algorithms such as the marching cubes are commonly used to develop a geometrical surface model of the tissue of interest. Stereolithography (STL) format is commonly used to store the geometrical model. An example of a 3D reconstructed lower-limb model is shown in Figure 1.10.

	ScanIP[7]	Amira[8]	Mimics[9]	SIP[10]
Input	Dicom, raw image, bmp, jpg	Dicom, raw image, bmp, jpg	Dicom, raw image, bmp, jpg	Dicom, bmp
Segmentation	Automatic, semi-automatic, manual	Automatic, semi-automatic, manual	Automatic, semi-automatic, manual	Semi-automatic
Main advantages	Manual segmentation for MRI-based complex biological tissues	Structured processing flow chart	Fast prototyping for CT-based implants, prosthesis, and orthosis	Flexible processing for FE meshing

Table 1.3. *Commercial and "home-made" image processing tools*

6 http://www.slicer.org/.

7 Commercial software: http://www.simpleware.com/software/scanip/.

8 Commercial software: http://www.vsg3d.com/amira/amira.

9 Commercial software: http://biomedical.materialise.com/mimics.

10 "Home-made" software: M.C. Ho Ba Tho ©INSERM 1991.

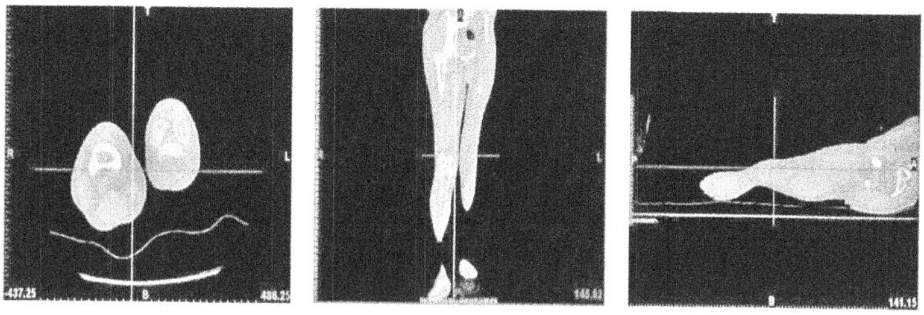

Figure 1.9. *Automatic segmentation of the lower limb structures*

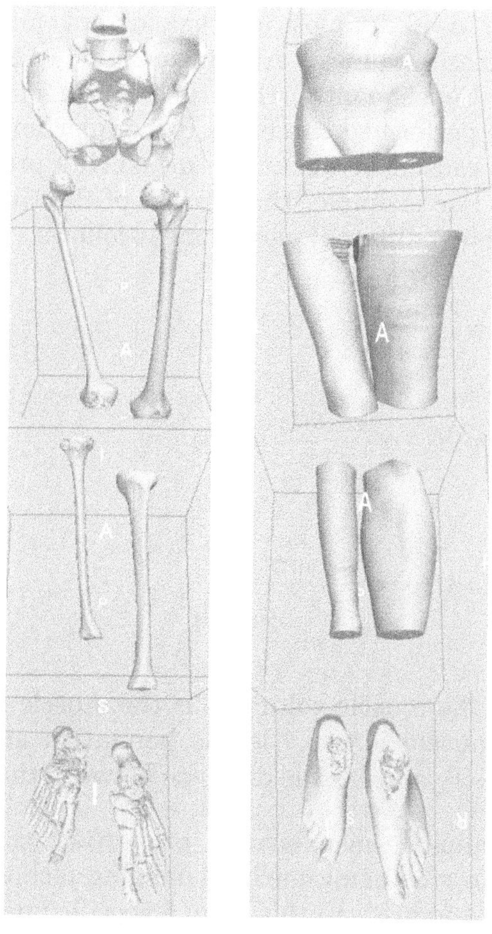

Figure 1.10. *3D reconstructed lower limb model*

c) *Computation of geometrical and anthropometrical properties*

Geometrical and anthropometrical properties of musculoskeletal tissues and structures are used as input data for the musculoskeletal model.

For the muscle tissue, common geometrical properties (e.g. physiological cross-sectional area (pCSA), volume, length and pennation angle) extracted from literature-based studies are used in most of the current musculoskeletal models. These values are measured using cadaveric specimens. Consequently, their use for *in vivo* simulation needs to be performed carefully. It is important to note that there is no consensus about the measuring process of these properties. For example, the muscle pCSA values could be computed using different equations as follows:

$$pCSA = \frac{V(\mathrm{cm}^3)}{l(cm)} \qquad [1.1]$$

$$pCSA = M(g) \times \frac{Cos\theta(°)}{\rho(\frac{g}{cm^3})} \times l(cm) \qquad [1.2]$$

$$pCSA = M(g) \times \frac{\sin\theta(°)}{\rho(\frac{g}{cm^3}) \times t(cm)} \qquad [1.3]$$

where V is the muscle volume, l is the muscle (fiber) length, M is the muscle mass, θ is the pennation angle, ρ is the muscle density and t is the distance between the tendons.

These muscle geometrical properties could now be measured *in vivo* using medical imaging techniques such as ultrasound and MRI. In this case, a specific image processing procedure needs to be applied.

For the BSIPs (i.e. anthropometrical properties), we could use regression tables (e.g. Dempster's [DEM 67] or Zatsiorsky's [ZAT 85] tables) to compute these values for a specific subject. However, these regression tables were established using cadaveric specimens. Subsequently, their use for *in vivo* studies could be made with great care. *In vivo* individualized BSIPs could be determined from medical imaging and anatomical axes using the parallel axis theorem. The ranges of values of BSIPs of a normal subject (male, 29 years old, 168 cm in height and 65 kg in weight) are given in Table 1.4.

Segment	Body Mass (%)	Ixx (kg m^2)	Iyy (kg m^2)	Ixx (kg m^2)
Right thigh	10.28 ± 0.27	0.06 ± 0.004	0.056 ± 0.004	0.021 ± 0.001
Left thigh	10.66 ± 0.3	0.063 ± 0.004	0.059 ± 0.004	0.023 ± 0.001
Right leg	4.5 ± 0.08	0.021 ± 0.001	0.021 ± 0.001	0.004 ± 0
Left leg	4.61 ± 0.09	0.022 ± 0.002	0.022 ± 0.001	0.004 ± 0
Right foot	1.1 ± 0.02	0.002 ± 0	0.001 ± 0	0.002 ± 0
Left foot	1.1 ± 0.03	0.002 ± 0	0 ± 0	0.002 ± 0

Table 1.4. *BSIP values of a normal subject*

d) *Constitutive laws*

Biological tissues and structures have complex behaviors. Modeling of biological tissues and structures needs to establish mathematical constitutive equations to describe these complex behaviors. Rigid multi-body musculoskeletal modeling requires constitutive laws for muscle, joint and contact behaviors.

For the muscle constitutive law, a Hill-type rheological model [HIL 38] is commonly used to define the force – length and force-velocity relationships [ZAJ 89]. A graphical representation of a Hill-type muscle model is shown in Figure 1.11. The Hill-based model has five input parameters: F_M^0 which is the peak isometric muscle force derived from the cross-sectional area of the muscle M; l_M^0 is the optimal muscle-fiber length, the length at which the muscle develops maximum force; α^0 is the optimal fiber length; l_T^S is the tendon slack length, the length at which tendons begin to transmit force when stretched; and $V_M^{\max}$ is the maximum contraction velocity of the muscle. The Hill-based model has three behavior curves: the tendon force-length curve, muscle force-length curve and muscle force-velocity curve as shown in Figure 1.12. F_M is the muscle force. l_T is the tendon length. l_M is the muscle length. V_M is the muscle contraction velocity. It is important to note that mechanical properties of the muscle (e.g. shear modulus) derived from magnetic resonance elastography (MRE) have been recently used to improve the behavior of the force – length relationship, leading to provide more accurate estimated muscle forces [BEN 13].

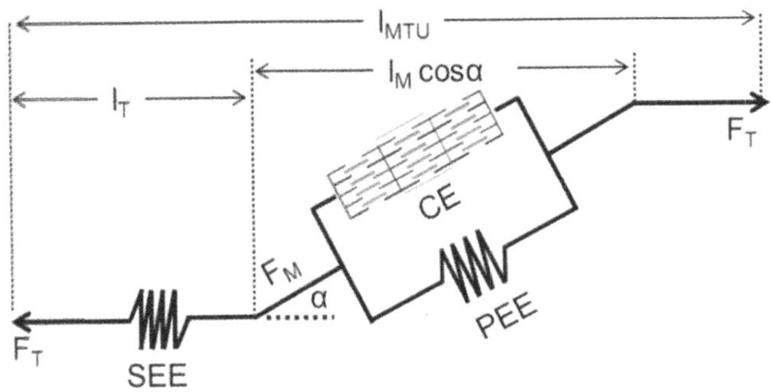

Figure 1.11. *Graphical representation of a Hill-type muscle model*

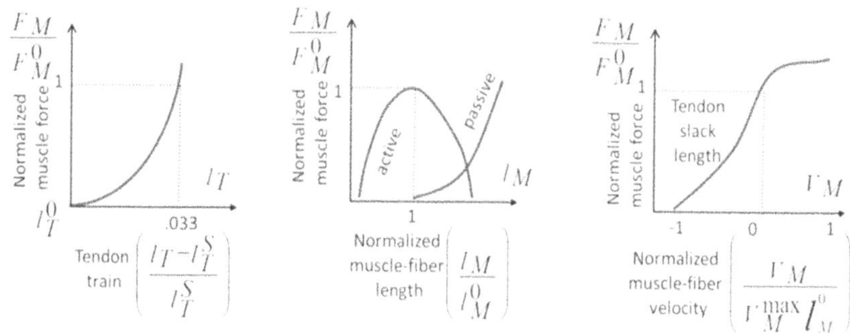

Figure 1.12. *Muscle behavior curves: tendon force–length curve, muscle force-length curve and muscle force–velocity curve*

For joint modeling, there are many types of joints such as ball-and-socket joints, hinge joints and spherical joints to model a biological joint. OpenSIM software provides a large range of joints (weld joint, pin joint, slider joint, ball-and-socket joint, ellipsoid joint, free joint and custom joint). The choice of the joint type depends on each specific biological joint and available kinematic data. For example, the hip joint is commonly modeled as a ball-and-socket joint. The knee joint is commonly modeled as a hinge joint. It is important to note that the current joint model includes only 3 degrees of freedom in rotation. Translations are always neglected. A joint coordinate system is commonly defined using the recommendation of the International Society of Biomechanics (ISB) [WU 02, WU 05]. An example of a viscoelastic joint law (Kelvin–Voigt model) is expressed below:

$$M = K \times \theta + C \times \dot{\theta} \tag{1.4}$$

where M is the joint moment, K and C are the joint stiffness and damping coefficients, respectively, and θ and $\dot{\theta}$ are the joint rotation angle and its angular velocity, respectively.

In addition to muscle and joint modeling, the interaction of the musculoskeletal model with the environment could be taken into consideration through contact modeling and constraints. The geometrical penetration principle is commonly used to model the contact between the model and the external structures. The deformation of rigid multi-body contact is generally computed using linear elastic theory [HUN 75, HER 82, PER 08].

e) *Rigid multi-body dynamics*

The *equations of motion* describing the dynamics of a musculoskeletal system are expressed as:

$$M(\mathrm{q})\ddot{\mathrm{q}}+\mathrm{C}(\mathrm{q},\dot{\mathrm{q}})+\mathrm{G}(\mathrm{q})+T_{MT}+E=0 \quad [1.5]$$

$$T_{MT}=R(q)\mathrm{F}^{MT} \quad [1.6]$$

where q is the joint angles set for n biological joints, $M(\mathrm{q})$ is the system mass ($n \times n$) matrix, $\mathrm{C}(\mathrm{q},\dot{\mathrm{q}})$ is the centrifugal and coriolis loading ($n \times 1$) matrix, $G(\mathrm{q})$ is the gravitational loading ($n \times 1$) matrix, T_{MT} is the muscular joint torques ($n \times 1$) matrix, $R(\mathrm{q})$ is the muscle moment arms ($n \times m$) matrix, F^{TM} is the muscle force ($m \times 1$) matrix and E is external forces (e.g. ground reaction forces). The muscle moment arms matrix could be calculated using the principle of virtual work. Thus, the moment arm of muscle j with respect to joint axis i is computed as follows:

$$R_{ij}(q)=-\frac{\partial L_j(q)}{\partial q_i} \quad [1.7]$$

where $L_j(q)$ is the length of muscle j.

Inverse dynamics is a computing algorithm which aims to estimate the torque of different biological joints. When the joint kinematics data (q) and external forces (E) are available, we could compute the muscle joint torques during motion by using the following inverse dynamics equation:

$$T_{MT} = M(q)\ddot{q} + C(q,\dot{q}) + G(q) + E \qquad [1.8]$$

Joint kinematics data are generally obtained by using motion capture systems such as VICON and motion analysis. These systems use high-resolution infrared cameras to track and obtain the trajectories of external skin-mounted markers using the optical principle. The configuration of skin-mounted markers depends on each specific application. For example, Davis's protocol [DAV 91] including 15 markers is commonly used for clinical gait analysis (as illustrated in Figure 1.13). Moreover, external forces such as foot–ground reaction forces could also be obtained synchronously using a force–plate system with kinematics data (as illustrated in Figure 1.14).

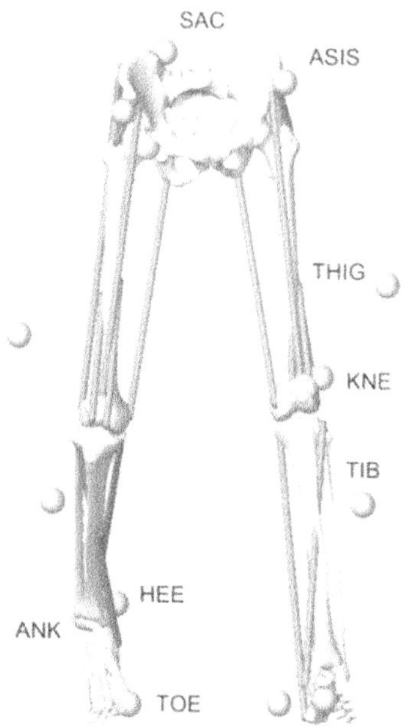

Figure 1.13. *Skin-mounted marker configuration using Davis's gait protocol*

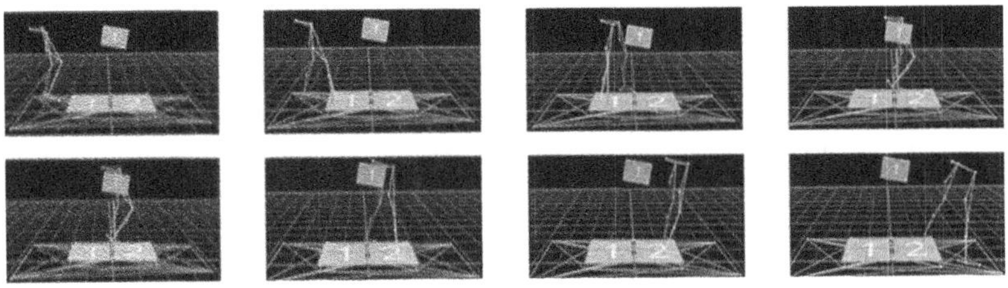

Figure 1.14. *Illustration of marker tracking and foot-ground reaction force acquisition during human gait*

Static optimization aims to estimate the muscle forces using computed net joint torque from inverse dynamics. From a mathematical point of view, the muscle force estimation problem is redundant because the number of unknown muscle forces is greater than the number of equations for all joints ($m > n$). Thus, to solve this problem, one of the widely used approaches is the inverse dynamics-based static optimization. The constitutive equations are expressed as follows:

$$\text{Minimize } F_{obj} \tag{1.9}$$

$$\text{Subject to } R(q)\mathrm{F}^{MT} = T^{MT} \tag{1.10}$$

$$0 \le \mathrm{F}^{MT} \le F_M^0 \tag{1.11}$$

where F_{obj} is a mono-objective or multi-objective function (e.g. it minimizes the total muscle forces $F_{obj} = \sum_{i=1}^{m} F_i^{MT}$).

An example of normal and pathological gait simulations using inverse dynamics and static optimization is shown in Figures 1.15 and 1.16, respectively.

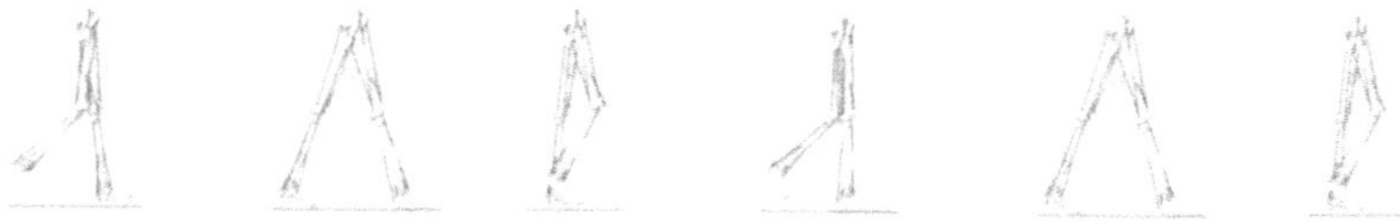

Figure 1.15. *Graphical representation of a normal gait simulation*

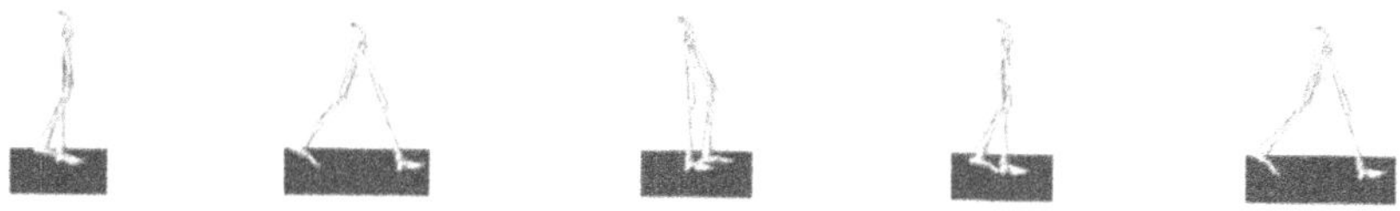

Figure 1.16. *Graphical representation of a pathological gait simulation*

Forward dynamics aims to compute the movement patterns using known muscle excitations or joint torques; the constitutive equation is expressed as follows:

$$\ddot{q} = M(q)^{-1}\left[C(q,\dot{q}) + G(q) + T_{MT} + E\right] \qquad [1.12]$$

f) *Partial validation*

The validation of the musculoskeletal model is a challenging issue for the modeling community. Due to the lack of experimental data for *in vivo* and non-invasive studies, simulation results are commonly compared to those reported in the literature. This approach is not completely convincing because the subject and modeling method of each study are different. Surface electromyography signals could be used to validate the muscle activation patterns. However, this technique also has limitations (e.g. position-dependent signals or a higher signal-to-noise ratio) leading to inaccurate signals. For the invasive studies, we could use the experimental joint force measured by an implanted device to validate the simulation results. In the same manner as the generic model, the full validation of the subject- or patient-specific model needs further investigation with new engineering approaches in order to provide more systematic validation before its clinical use [DAO 09].

1.3. Challenges and perspectives of rigid multi-body musculoskeletal models

The rigid multi-body dynamics principle allows human motion to be described and analyzed in qualitative and quantitative manners. By coupling the motion laws with biological laws (e.g. the muscle rheological law), the kinematics, kinetics and muscle force data could be derived and analyzed. However, the understating of muscle force generation capacity is still a challenging issue in research. From a clinical point of view, such muscle force data is of great importance to depict the etiology of a pathological gait and to perform the objective diagnosis and evaluation of the treatment outcome. To reach this clinic-oriented objective, current musculoskeletal modeling needs to be improved in the following areas:

– *Biological joint*: sliding translations need to be integrated into current joint models. This improvement will allow accurate joint behavior to be modeled and simulated, especially in the case of patients with irregular geometries. Moreover, there is no consensus about the choice of the joint type for a specific biological joint. Consequently, a new approach needs to be investigated to allow a biological joint to be modeled in a subject- or patient-specific manner.

– *Biological skeletal muscle*: geometrical representation of the skeletal muscle needs to be improved. The line-of-action representation and one-dimensional (1D) rheological model (e.g. Hill-type model) are very limited in modeling the real biological skeletal muscle. Medical imaging techniques such as MRI could be used to create 3D geometries of the muscle of interest. However, the time-consuming characteristic of MRI data acquisition and processing needs to be overcome. The Hill-type model is commonly used in the musculoskeletal models. This model integrates force–length and force–velocity relationships to describe the muscle contraction behaviors. Moreover, literature-based values are commonly used leading to inaccurate muscle force estimation. As a result, experimental imaging techniques such as ultrasound and MRE [BEN 13] could be used to compute the individualized values for muscle properties. Furthermore, mechanical muscle properties need to be integrated into muscle models to accurately simulate the muscle behavior, especially in the case of patients with muscle diseases. Moreover, individualized mono- or multi-objective functions will avoid the dilemma of the right choice of an appropriate function for a specific case. An EMG signal could be integrated to develop an EMG-driven musculoskeletal model in order to better describe muscle activities and behaviors. Another challenging issue will be the integration of 3D muscle constitutive laws [BLE 05, TAN 09] to better describe the muscle behavior. Furthermore, the muscle–bone penetration problem (as illustrated in Figure 1.17) needs to be improved.

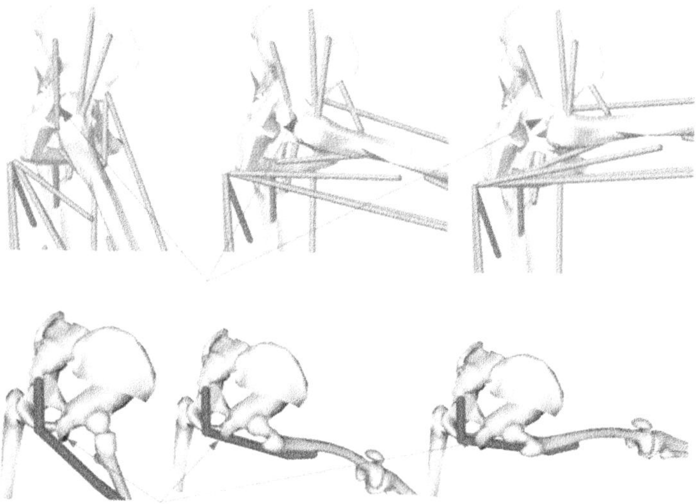

Figure 1.17. *Problem of muscle–bone contact penetration: muscle line representation (top) and muscle wrapping representation (bottom)*

– *Nervous system*: the development of a neuromusculoskeletal model will integrate neural control command into a numerical model. In this case, innovative methods and techniques need to be developed to acquire the nervous signals and then use them to accurately simulate the normal and pathological states of the human body.

– *Validation*: experimental validation of the simulation results of the musculoskeletal model is a very challenging issue. A multi-modal validation approach needs to be investigated. The accuracy of the EMG signal needs to be improved. Then, this signal could be used to validate the muscle contraction patterns or to serve as input data into the musculoskeletal model. Novel methods and techniques need to be developed to provide new *in vivo* data for validation purpose.

– *Clinical applications*: the choice of the pertinent parameters for clinical purpose is also a challenge. To make the musculoskeletal model applicable to a real clinical context, this choice is crucial. Collaboration among model developers, clinicians and patients needs to be performed closely to target the appropriate diagnosis model for a specific clinical case. Then, the model could be used to evaluate the outcome of clinical treatments.

– *Biomechanical data uncertainty*: the impact of uncertainties of experimental data on the numerical output responses has been challenged in the last decade [VAL 03, DOW 06, RIE 08, DHA 10, DAO 12]. However, most of the biomechanical research studies concentrated on the quantification of such an impact by using a traditional variation and perturbation approach [DAO 12, NAG 00, SCO 06]. The limitation of this approach deals with the not obvious choice of the true range of values of a parameter of interest for performing the sensitivity analysis. In addition, the resolution algorithm for a large range of values is computationally intensive or impractical in some cases. Moreover, uncertainty sources and types cannot be identified and modeled. Furthermore, dependence (e.g. statistical, physical or biological dependence) between parameters cannot be taken into consideration. Chapter 2 will present basic theoretical and practical concepts focusing on these challenging topics.

This chapter provides an introduction of the biomechanics field of study and its potential application range. Current knowledge of biomechanics of the musculoskeletal system is presented. The focus is on the development of a rigid multi-body musculoskeletal model. Discussion of limitations and possible improvements of this kind of biomechanics modeling approach provides us with a clear insight into the current models and how to make the model applicable in a real clinical context.

1.4. Summary

– Biomechanics: study of living systems using mechanical engineering methods, techniques and theories.

– Biomechanics applications: virtual muscle-tendon surgery, optimal design of biomedical materials and devices, assessment of gait abnormalities, computer-aided orbital and maxillofacial surgery, detection and prediction of preterm deliveries, sportive performance analysis, electrical energy harvesting, early diagnosis of degenerated IVD, etc.

– Musculoskeletal model: numerical representation of the interaction between bones, joints, muscles, ligaments and their attachments to the bones.

– Generic musculoskeletal model: parameterized and scaling process for a specific subject or patient.

– Subject– or patient–specific musculoskeletal model: individualized model with subject– or patient–specific data properties.

– Modeling assumptions need to be defined due to limited knowledge and/or difficult technical implementation.

– Modeling challenges: biological joints, 3D realistic muscle models, neuromusculoskeletal model, model validation, clinical application and biomechanical data uncertainty consideration.

1.5. Bibliography

[ARN 01] ARNOLD A.S., BLEMKER S.S., DELP S.L., "Evaluation of a deformable musculoskeletal model: application to planning muscle-tendon surgeries for crouch gait", *Annals of Biomedical Engineering*, vol. 29, pp. 1–11, 2001.

[ARN 04] ARNOLD A.S., DELP S.L., "The role of musculoskeletal models in patient assessment and treatment", in GAGE J.R. (ed.), *Treatment of Gait Problems in Cerebral Palsy*, Cambridge Press, 2004.

[ARN 05] ARNOLD A.S., DELP S.L. "Computer modeling of gait abnormalities in cerebral palsy: application to treatment planning", *Theoretical Issues in Ergonomics Science*, vol. 6, pp. 305–312, 2005.

[ASA 02] ASAKAWA D.S., BLEMKER S.S., GOLD G.E., *et al.* "*In vivo* motion of the rectus femoris muscle after tendon transfer surgery", *Journal of Biomechanics*, vol. 35, pp. 1029–1037, 2002.

[BEN 13] BENSAMOUN S.F., DAO T.T., CHARLEUX F., *et al.*, "Estimation of muscle force derived from *in vivo* MR elastography tests: a preliminary study", *Journal of Musculoskeletal Research*, vol. 16, no. 3, pp. 1350015–1350025, 2013.

[BLE 05] BLEMKER S.S., PINSKY P.M., DELP S.L., "A 3D model of muscle reveals the causes of nonuniform strains in the biceps brachii", *Journal of Biomechanics*, vol. 38, pp. 657–665, 2005.

[BUC 07] BUCHAILLARD S., BRIX M., PERRIER P., *et al.*, "Simulations of the consequences of tongue surgery on tongue mobility: implications for speech production in post-surgery conditions", *International Journal of Medical Robotics and Computer Assisted Surgery*, vol. 3, no. 3, pp. 252–261, 2007.

[BUR 06] BURGESS D.J., NAUGHTON G., NORTON K.I., "Profile of movement demands of national football players in Australia", *Journal of Science and Medicine in Sport*, vol. 9, no. 4, pp. 334–341, 2006.

[CAR 09] CARPENTIER B., GAUTIER A., LEGALLAIS C. "Artificial and bioartificial liver devices: present and future", *Gut*, vol. 58, pp. 1690–1702, 2009.

[CHA 97] CHANG C.W., LEE S.Y., "A video information system for sport motion analysis", *Journal of Visual Languages & Computing*, vol. 8, no. 3, pp. 265–287, 1997.

[DAO 09] DAO T.T., Modeling of musculoskeletal system of the lower limbs: biomechanical model vs. meta model, PhD Thesis, University of Technology of Compiègne, France, 2009.

[DAO 12] DAO T.T., MARIN F., POULETAUT P., *et al.*, "Estimation of accuracy of patient specific musculoskeletal modeling: case study on a post-polio residual paralysis subject", *Computer Method in Biomechanics and Biomedical Engineering*, vol. 15, no. 7, pp. 745–751, 2012.

[DAO 13] DAO T.T., POULETAUT P., ROBERT L., *et al.*, "Quantitative analysis of annulus fibrosus and nucleus pulposus derived from T2 mapping, diffusion-weighted and diffusion tensor MR imaging", *Computer Methods in Biomechanics and Biomedical Engineering: Imaging & Visualization*, vol. 1, 2013.

[DAV 91] DAVIS R.B., OUNPUU S., TYBURSKI D., *et al.*, "A gait analysis data collection and reduction technique", *Human Movement Science*, vol. 10, pp. 575–587, 1991.

[DEL 97] DELP S.L., LOAN J.P., BASDOGAN C., *et al.*, "Surgical simulation: an emerging technology for emergency medical training", *Presence: Teleoperators and Virtual Environments*, vol. 6, pp. 147–159, 1997.

[DEM 67] DEMPSTER W.T., GAUGHRAN G.R.L., "Properties of body segments based on size and weight", *American Journal of Anatomy*, vol. 120, no. 1, pp. 33–54, 1967.

[DHA 10] DHAHER Y.Y., KWON T.H., BARRY M., "The effect of connective tissue material uncertainties on knee joint mechanics under isolated loading conditions", *Journal of Biomechanics*, vol. 43, no. 16, pp. 3118–3125, 2010.

[DIA 09] DIAB M., MARQUE C., KHALIL M. "An unsupervised classification method of uterine electromyography signals classification for detection of preterm deliveries", *Journal of Obstetrics and Gynaecology Research*, vol. 35, no. 1, pp. 9–19, 2009.

[DON 08] DONELAN J.M., LI Q., NAING V., *et al.*, "Biomechanical energy harvesting: Generating electricity during walking with minimal user effort", *Science*, vol. 319, pp. 807–810, 2008.

[DOW 06] DOWLING J.J., DURKIN J.L., ANDREWS D.M., "The uncertainty of the pendulum method for the determination of the moment of inertia", *Medical Engineering & Physics*, vol. 28, no. 8, pp. 837–841, 2006.

[HAS 10] HASSAN M., TERRIEN J., KARLSSON B., *et al.*, "Interactions between uterine emg at different sites investigated using wavelet analysis: comparison of pregnancy and labor contractions", *EURASIP Journal on Advances in Signal Processing*, vol. 2010, pp. 1–9, 2010.

[HAT 74] HATZE H., "The meaning of the term biomechanics", *Journal of Biomechanics*, vol. 7, pp. 189–190, 1974.

[HAU 04] HAUGHTON V., "Medical imaging of intervertebral disc degeneration: current status of imaging", *Spine*, vol. 29, pp. 2751–2756, 2004.

[HER 82] HERTZ H., "On the contact of elastic solids", *Journal Für Die Reine und Angewandte Mathematik*, vol. 92, pp. 156–171, 1882.

[HIL 38] HILL A.V. "The heat of shortening and dynamics constants of muscles", *Proc. R. Soc. Lond. B*, vol. 126, no. 843, pp. 136–195, 1938.

[HUN 75] HUNT K.H., CROSSLEY F.R.E., "Coefficient of restitution interpreted as damping in vibroimpact", *ASME Journal of Applied Mechanics*, vol. 42, pp. 440–445, 1975.

[KEY 65] KEYNES R.D., "Biological instrumentation", *Nature*, vol. 19, p. 60, 1965.

[KIT 02] KITANO H., "Computational systems biology", *Nature*, vol. 420, pp. 206–210, 2002.

[KUO 05] KUO A.D., "Harvesting energy by improving the economy of human walking", *Science*, vol. 309, pp. 1686–1687, 2005.

[LUB 05] LUBOZ V., CHABANAS M., SWIDER P., *et al.*, "Orbital and maxillofacial computer aided surgery: patient-specific finite element models to predict surgical outcomes", *Computer Methods in Biomechanics & Biomedical Engineering*, vol. 8, no. 4, pp. 259–265, 2005.

[NAG 00] NAGANO A., GERRITSEN K.G.M., FUKASHIRO S., "A sensitivity analysis of the calculation of mechanical output through inverse dynamics: a computer simulation study". *Journal of Biomechanics*, vol. 33, no. 10, pp. 1313–1318, 2000.

[PER 02] PERIE D., SALES DE GAUZY J., HO BA THO M.C., "Biomechanical evaluation of Cheneau-Toulouse-Munster brace in the treatment of scoliosis using optimisation approach and finite element method", *Med Biol Eng Comput*, vol. 40, no. 3, pp. 296–301, 2002.

[PER 08] PEREZ-GONZALEZ A., FENOLLOSA-ESTEVE C., SANCHO-BRU J.L., *et al.*, "A modified elastic foundation contact model for application in 3d models of the prosthetic knee", *Medical Engineering & Physics*, vol. 30, no. 3, pp. 387–398, 2008.

[REI 02] REIF J.H., "Computing: successes and challenges", *Science*, vol. 19, pp. 478–479, 2002.

[RIE 97] RIEWALD S.A., DELP S.L. "The action of the rectus femoris muscle following distal tendon transfer: does it generate a knee flexion moment?", *Developmental Medicine and Child Neurology*, vol. 39, pp. 99–105, 1997.

[RIE 08] RIEMER R., HSIAO-WECKSLER E.T., ZHANG X., "Uncertainties in inverse dynamics solutions: a comprehensive analysis and an application to gait", *Gait & Posture*, vol. 27, no. 4, pp. 578–588, 2008.

[SCO 06] SCOVIL C.Y., RONSKY J.L., "Sensitivity of a Hill-based muscle model to perturbations in model parameters", *Journal of Biomechanics*, vol. 39, no. 11, pp. 2055–2063, 2006.

[SHA 01] SHADISH W.R., THOMAS D.C., DONALD T.C., *Experimental and Quasi-experimental Designs for Generalized Causal Inference*, Houghton Mifflin, Boston, 2001.

[SLO 98] SLOTEN J.V., HO BA THO M.C., VERDONCK P., "Applications of computer modelling for the design of orthopaedic, dental and cardiovascular biomaterials", *Proc Inst Mech Eng Part H*, vol. 212, no. 6, pp. 489–500, 1998.

[SPE 05] SPENCER M., RECHICHI C., LAWRENCE S., *et al.*, "Time-motion analysis of elite field hockey during several games in succession: A tournament scenario", *Journal of Science and Medicine in Sport*, vol. 8, no. 4, pp. 382–391, 2005.

[TAN 09] TANG C.Y., ZHANG G., TSUI C.P., "A 3D skeletal muscle model coupled with active contraction of muscle fibres and hyperelastic behavior", *Journal of Biomechanics*, vol. 42, pp. 865–872, 2009.

[VAL 03] VALERO-CUEVAS F.J., JOHANSON M.E., TOWLES J.D., "Towards a realistic biomechanical model of the thumb: the choice of kinematic description may be more critical than the solution method or the variability/uncertainty of musculoskeletal parameters", *Journal of Biomechanics*, vol. 36, no. 7, pp. 1019–1030, 2003.

[VEN 06] VENTURA B.D., LEMERLE C., MICHALODIMITRAKIS K., *et al.*, "From *in vivo* to in *silico* biology and back", *Nature*, vol. 443, pp. 527–533, 2006.

[WIN 11] WINTER D.A., *Biomechanics and Motor Control of Human Movement*, John Wiley & Sons, 2011.

[WU 02] WU G., SIEGLER S., ALLARD P., *et al.*, "ISB recommendation on definitions of joint coordinate system of various joints for the reporting of human joint motion – part I: ankle, hip, and spine", *Journal of Biomechanics*, vol. 35, no. 4, pp. 543–548, 2002.

[WU 05] WU G., VAN DER HELM F.C., VEEGER H.E., *et al.*, "International Society of Biomechanics. ISB recommendation on definitions of joint coordinate systems of various joints for the reporting of human joint motion – part II: shoulder, elbow, wrist and hand", *Journal of Biomechanics*, vol. 38, no. 5, pp. 981–992, 2005.

[ZAJ 89] ZAJAC F.E., "Muscle and tendon: properties, models, scaling, and application to biomechanics and motor control", *Crit Rev Biomed Eng*, vol. 17, no. 4, pp. 359–411, 1989.

[ZAT 85] ZATSIORSKY V.M., SELUYANOV V.N., "Estimation of the mass and inertia characteristics of the human body by means of the best predictive regression equations", *Biomechanics*, vol. 9, pp. 233–239, 1985.

2

Modeling of Biomechanical Data Uncertainty

Experimental investigation combined with numerical simulations is commonly used for solving multiphysical problems. In the field of biomechanics which aims to understand the mechanics of living systems, the main difficulty is to provide the experimental data reflecting the multiphysical behavior of the systems of interest. These experimental data are used as input data for numerical simulations to quantify output responses through physical and/or biological laws expressed by constitutive mathematical equations. Moreover, uncertainties on the experimental available data exist as human variability, measuring protocols and numerical processing. This chapter describes the fundamental and conceptual aspects of the data uncertainty modeling in biomechanics. Different biomechanics data types and related parameters such as physiological, morphological, mechanical, and kinematics and kinetics properties and their uncertainty sources (e.g. experimental and numerical) are identified and introduced. Modeling approaches based on the types and representations of uncertainty are presented. Finally, an example of the propagation of data uncertainty and decision-making through a numerical model are presented and discussed.

2.1. Introduction of biomechanical data and their uncertainties

2.1.1. *Biomechanical data*

Biomechanical data are essential for any experimental or modeling studies. Biomechanical data are commonly used to describe the anatomical, mechanical and functional behaviors of biological tissues and systems. Most biomechanical data are classified into four groups (see Figure 2.1): (1) physiological properties, (2) morphological properties, (3) mechanical properties and (4) motion properties including spatiotemporal, kinematic, kinetic and EMG properties. Let us discuss them each briefly.

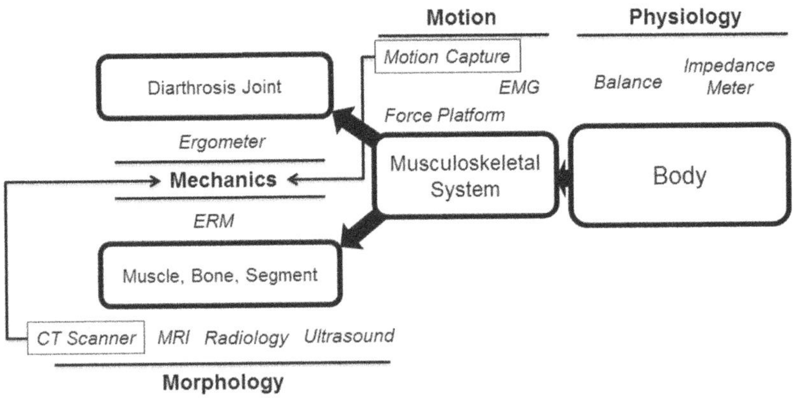

Figure 2.1. *Overview of biomechanical data*

Physiological properties relate to personal information at the whole body level of a subject under investigation, such as age, body height, body weight, body fat and body mass index (BMI). Standard measuring devices and techniques such as scales, height measuring devices or impedance meters are commonly used to obtain these data. The BDI is calculated using the body weight (W), body height (H) and the following formula:

$$BDI = \frac{\mathrm{W(kg)}}{\mathrm{H(m)}^2} \qquad [2.1]$$

Morphological properties deal with the geometrical (e.g. shape or form) and structural (e.g. organization) characteristics of biological tissues such as bone, muscle or segmental bodies (e.g. the thigh and leg). Medical imaging techniques such as ultrasound, computed tomography, MRI or radiography are commonly used to obtain these data. Some examples of morphological properties are given in Table 2.1.

Anatomy	Properties
Bone	Length (femur, tibia and foot) (cm), volume (femur, tibia and foot) (cm^3), femoral anteversion angle (°), tibial torsion angle (°), migration index (MI)
Muscle	Muscle length (cm), fiber length (cm), volume (cm^3), fiber pennation angle (°), pCSAs (cm^2)
Segmental body (e.g. thigh)	Segmental mass (kg), positions of center of mass, segmental moment of inertia (kg m^2), leg length (m), thigh length (m), ankle width (m)

Table 2.1. *Examples of morphological properties*

Mechanical properties describe the behavioral reaction of complex biological materials such as bone or muscle under an applied load. These properties depend mainly on the multiscale compositions and structures of biological materials. Some examples of mechanical properties are given in Table 2.2.

Motion properties characterize the functional behaviors of the biological systems during motions such as daily and sports activities. Motion analysis properties include whole body (e.g. spatiotemporal properties) or joint behavioral properties (e.g. kinematics and kinetics) and muscle activity through EMG properties. Most kinematic, kinetic and EMG data are expressed by temporal waveforms. The knee joint angle during walking is an example of a kinematic property. Other examples are presented in Table 2.3.

Anatomy and structure	Properties
Bone	Young's modulus (E) (kPa), Poisson ratio, shear modulus (kPa), rupture (σ) (Mpa)
Muscle	Shear modulus (G) (kPa), attenuation coefficient (m^{-1}), velocity-dependent spasticity ($\mu V \times \deg^{-1}$)
Joint	Stiffness ($Nm \times \deg^{-1}$), slack angle (°)

Table 2.2. *Examples of mechanical properties*

Anatomy and structure	Subtype	Properties
Whole body	Spatiotemporal	Step length (cm), step width (cm), cadence (steps per minute), speed (cm/s), stance time (%), swing time (%), double-support time (%), single-support time (%), energy consumption
Joint	Kinematics	Angles (hip, knee, ankle) (°), velocity (m/s), acceleration (m/s²), Gillette gait index, gait deviation index, gait profile score
Joint	Kinetics	Moments (hip, knee, ankle) (Nm), reaction forces (hip, knee, ankle) (N), ground reaction forces (N)
Muscle		Muscle strain (mm) and muscle forces (N)
Muscle	EMG	Maximal and minimal myoelectrical amplitudes (mV), EMG onset, integrated EMG (mV), root mean square (mV), mean frequency (Hz)

Table 2.3. *Examples of kinematic, kinetic and EMG properties*

2.1.2. *Measuring chains of biomechanical data*

The measuring process of biomechanical data is shown in Figure 2.2. The acquisition of physiological data does not need specific processing. These properties could be measured and obtained directly from measuring devices. There are two main experimental techniques widely used for the acquisition of morphological, mechanical and motion analysis properties. The first technique is medical imaging and the second technique is 3D motion capture. Thus, from a raw image or temporal data, specific imaging and signal processing needs to be performed to compute useful data.

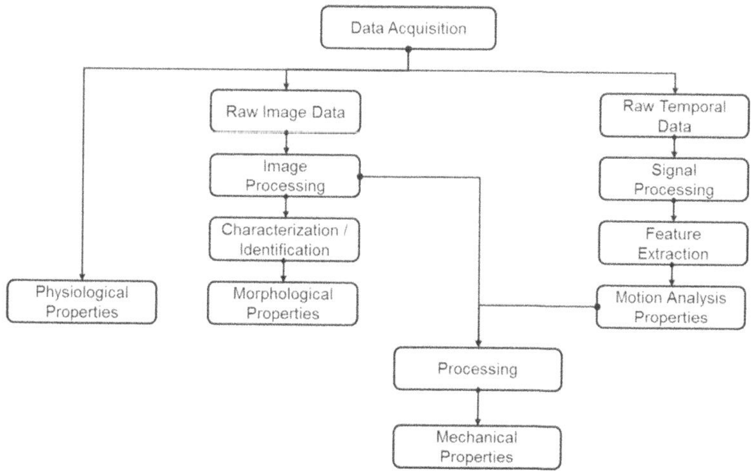

Figure 2.2. *Measuring chain of biomechanical data*

Morphological properties are commonly calculated from image-based data. Raw image data from medical imaging systems are stored in DICOM format. To compute the 3D morphological properties such as volume and the length of the biological tissue of interest, imaging processing methods such as image filtering, segmentation and 3D reconstruction are commonly used. The image filtering aims to improve the quality of the raw image by deleting some undesirable image characteristics (e.g. intensity variation, illumination variation and poor contrast). The segmentation aims to assign pixels/voxels to a region of interest (e.g. biological tissue). There are three common methods used for this: manual, automatic and semi-automatic segmentations.

Automatic segmentation is generally applied to CT images. Manual segmentation is commonly used for MRI images. Semi-automatic segmentation could be used for both CT and MRI images. 3D reconstruction aims to create 3D representation of the segmented region of interest. The marching cube algorithm is a commonly used reconstruction method. Reconstructed geometries are stored in STL for further calculation of morphological properties.

Motion properties are computed using signal-processing techniques from a raw signal stored in C3D format (i.e. the common file format for storing 3D and analog information of motion data obtained from motion capture systems). Motion analysis data are commonly expressed by temporal waveforms. These temporal data are of great importance in the biomechanical analysis of human locomotion. This describes the time-dependent events at regular time intervals. Human gait data ranging from stance phase (0–60%) to swing phase (60–100%) is a typical example of time series data. To facilitate the implementation as well as to reduce the time processing without loss of significant information, a vector of useful extracted features is an appropriate representation of the time series data.

Feature selection derived from kinematic data: kinematics data include joint angle amplitudes at the hip, knee and ankle over gait cycle. Maximal extension amplitude of the hip joint ($H_{maxStExt}$) and its gait cycle percent ($H_{timeMaxStExt}$) in the stance phase are considered as useful properties (Figure 2.3). At the knee joint, maximal flexion amplitudes ($K_{maxStFlex}$, $K_{maxSwFlex}$) and their respective gait cycle percents ($K_{timeMaxStFlex}$, $K_{timeMaxSwFlex}$) in the stance phase and swing phase, respectively, are selected as useful features (Figure 2.3). Maximal extension amplitude of the knee joint ($K_{maxStExt}$) and its gait cycle percent ($K_{timeStMaxExt}$) in the stance phase were also selected (Figure 2.4). For the ankle joint, maximal plantar flexion and dorsiflexion amplitudes

($A_{maxStPlFlex}$, $A_{maxStDsFlex}$) and their respective gait cycle percents ($A_{timeMaxStPlFlex}$, $A_{timeMaxStDsFlex}$) in the stance phase are selected (Figure 2.4). Maximal plantar flexion and dorsiflexion amplitudes of the ankle joint ($A_{maxSwPlFlex}$, $A_{maxStDsFlex}$) and their respective gait cycle percents ($A_{timeMaxSwPlFle}$, $A_{timeMaxSwDsFlex}$) in the swing phase are also selected (Figure 2.5).

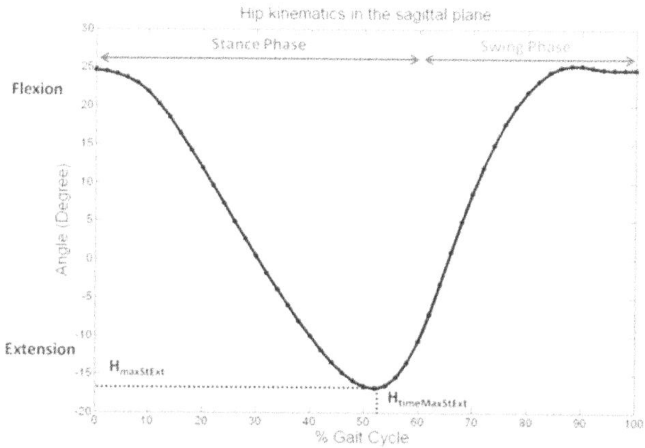

Figure 2.3. *Feature selection from hip kinematics in the sagittal plane*

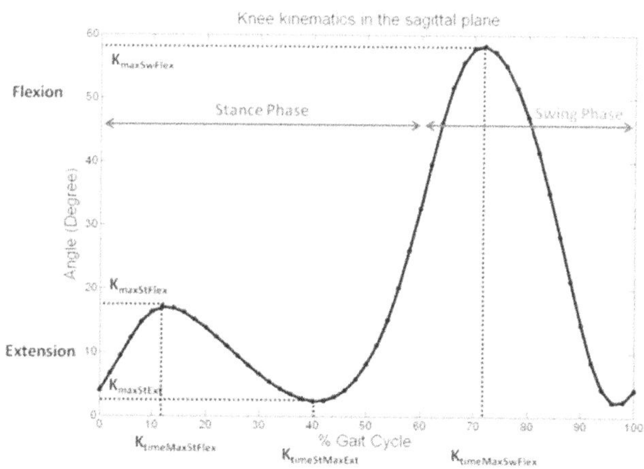

Figure 2.4. *Feature selection from knee kinematics in the sagittal plane*

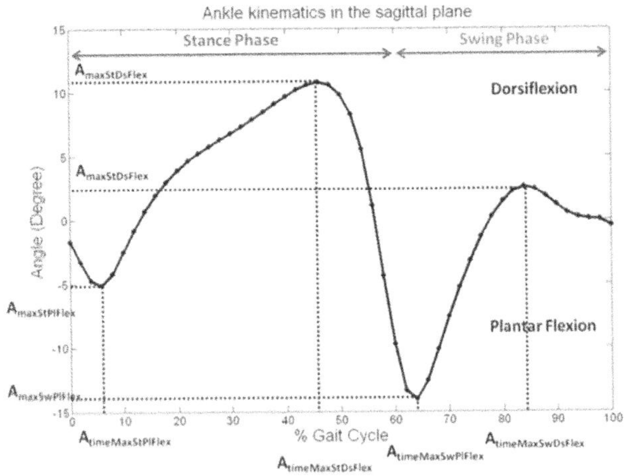

Figure 2.5. *Feature selection from ankle kinematics in the sagittal plane*

Feature selection derived from kinetics data: kinetics data include foot-ground reaction forces over the stance phase of the gait cycle. The vector of feature from kinetics data consists of the ground reaction forces (GRF_{to}, GRF_{hs}) and their respective contact times (T_{to}, T_{hs}) at the toe-off and heel strike phases are selected (Figure 2.6).

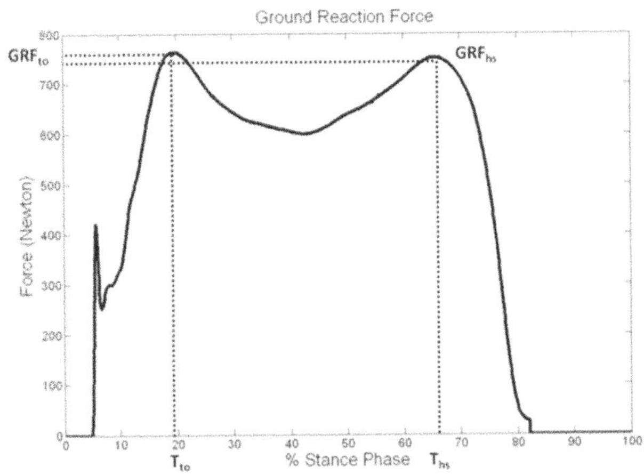

Figure 2.6. *Feature selection from kinetics data in the sagittal plane*

Feature selection derived from kinetics data: feature selection derived from an EMG signal – EMG data reflect the muscle myoelectrical activities over gait cycle. The vector of features from rectified EMG data in temporal domain consists of the maximal amplitude peak ($EMG_{maxPeak}$), average rectified voltage (EMG_{arv}) and EMG onset (i.e. % of gait cycle) (EMG_{onset}) (Figure 2.7).

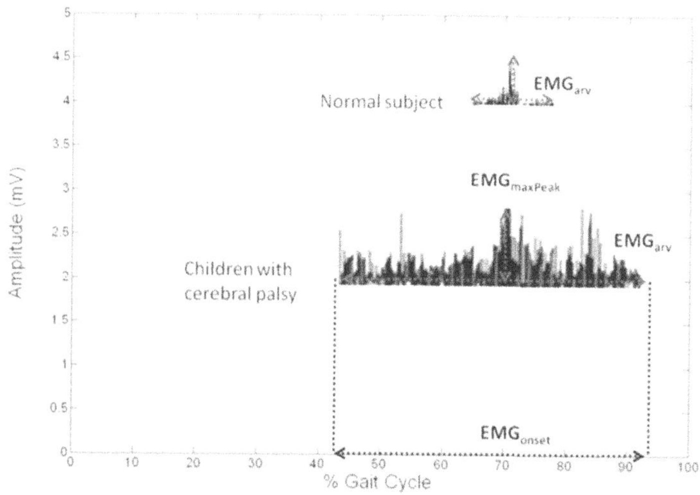

Figure 2.7. *Feature selection from rectified EMG signal: illustrations on normal subject and children with cerebral palsy*

Mechanical properties are computed data extracted from medical imaging or motion capture systems. For example, muscle shear modulus is characterized using MRE [BEN 06, RIN 07]. Another example is joint stiffness, which is computed from the joint kinematics [BAR 11].

2.1.3. *Data uncertainty*

Uncertainty is the lack of certainty. Data uncertainty is the lack of certainty about the right value of data. Data uncertainty is of two distinct types. The first type is the random uncertainty, which regards the variability of a parameter of interest under its systematic (intrinsic) functional and behavioral variations. The second type is the epistemic uncertainty, which deals with the lack of knowledge, the conflicting evidence, or the ignorance about a parameter of interest and its measuring protocol. Note that the random uncertainty is irreversible while the epistemic uncertainty is reversible.

EXAMPLE 2.1.– A property having a range of values (mean ± standard deviation) calculated from a number of repeated experiments is an example of random uncertainty.

EXAMPLE 2.2.– A property value that could be measured using different experimental techniques or computation methods due to the lack of knowledge about the right solution is an example of epistemic uncertainty.

2.1.4. *Biomechanical data uncertainty types and sources*

Biomechanical data (e.g. physiological, morphological, mechanical, kinematic, kinetic or EMG properties) are subject to random uncertainty regarding the measured range of values (i.e. intrinsic intrasubject variability and intersubject variability) of a parameter of interest. The intrinsic intrasubject variability concerns the repeatability and the reproducibility errors while the intersubject variability is due to the data obtained from different protocols or population races/origins or experimental techniques. Moreover, these biomechanical data are subject to the epistemic uncertainty dealing with the accuracy level of the measuring protocol including experimental and numerical processes. Furthermore, these data arise from multiple data sources (i.e. research studies) for one parameter of interest. In addition, biomechanical parameters can be either dependent (e.g. statistical, physical or biological dependence) or independent.

EXAMPLE 2.3.– An example of biomechanical random data uncertainty is that the physiological cross-sectional area of the rectus femoris muscle measured from 21 cadaveric specimens is 13.5 ± 5 cm^2 [WAR 09].

EXAMPLE 2.4.– An example of biomechanical epistemic data uncertainty is that the physiological cross-sectional area of the skeletal muscle could be calculated using one of the three formulas below:

$$pCSA = \frac{V(\mathrm{cm}^3)}{l(cm)} \quad [2.2]$$

$$pCSA = M(g) \times \frac{Cos\theta(°)}{\rho(\frac{g}{cm^3})} \times l(cm) \quad [2.3]$$

$$pCSA = M(g) \times \frac{\sin\theta(°)}{\rho(\frac{g}{cm^3}) \times t(cm)} \quad [2.4]$$

where V is the muscle volume, l is the muscle (fiber) length, M is the muscle mass, θ is the pennation angle, ρ is the muscle density and t is the distance between the tendons.

The sources of biomechanical data uncertainty could come from different aspects such as experimental measuring protocols and numerical processing. These sources could be divided into three categories: sources from direct measurements, sources from image-based measurements and sources from motion-based measurements. All these sources are given in Tables 2.4–2.6. Thus, physiological properties have uncertainty sources from direct measurements. Morphological properties have uncertainty sources from image-based measurements. Motion analysis properties have uncertainty sources from motion-based measurements. Mechanical properties have uncertainty sources from either image-based or motion-based measurements. This depends on the type of the properties used for feature calculation process.

Sources	Type	Uncertainty impact
Subject variability	Random	Repeatability and reproducibility (inter- and intraoperators, inter- and intratrials)
Measuring device error	Random	Reported range of errors for each specific device
Limited quantity of subjects/trials	Epistemic	Range of values

Table 2.4. *Uncertainty sources from direct measurements*

Sources	Type	Uncertainty impact
Subject variability	Random	Repeatability and reproducibility (inter- and intraoperators, inter- and intratrials)
Measuring protocol setup (parameters)	Epistemic	Range of values
Measuring device error	Random	Reported range of errors for each specific device
Data acquisition (position setup, movement artifact, tissue complexity and environmental effects)	Random	Range of values
Data processing (filtering, segmentation and reconstruction)	Random	Range of values
Limited quantity of subjects/trials	Epistemic	Range of values

Table 2.5. *Uncertainty sources from image-based measurements*

Sources	Type	Uncertainty impact
Subject variability	Random	Repeatability and reproducibility (inter- and intraoperators, inter- and intratrials)
Measuring protocol setup (parameters)	Epistemic	Range of values
Measuring device error	Random	Reported range of errors for each specific device
Data acquisition (position setup, skin-mounted marker setup, movement artifact, soft tissue artifacts and electronic noises)	Random	Range of values
Data processing (properties extraction, filter accuracy (smoothing for missing data)	Random	Range of values
Limited quantity of subjects/trials	Epistemic	Range of values

Table 2.6. *Uncertainty sources from motion-based measurements*

2.2. Biomechanical data uncertainty modeling

2.2.1. *Uncertainty representation*

2.2.1.1. *Classical probability functions*

Classical probability functions such as the probability density function (PDF) and cumulative distribution function (CDF) are common uncertainty representation formalisms [POT 12]. These functions are governed by the following equations:

$$PDF_R = CDR_R : V \to [0,1] \qquad [2.5]$$

$$PDF_R(x) = \Pr(R = x) \sum_{x \in V} PDF_R(x) = 1 \qquad [2.6]$$

$$CDF_R(x) = \Pr(R \leq x) \qquad [2.7]$$

EXAMPLE 2.5.– PDF and CDF of the volume of the rectus femoris muscle are shown in Figures 2.8 and 2.9.

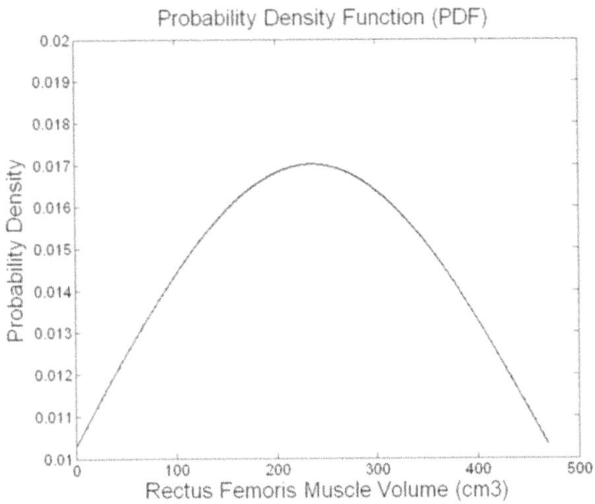

Figure 2.8. *Probability density function of the volume of the rectus femoris muscle*

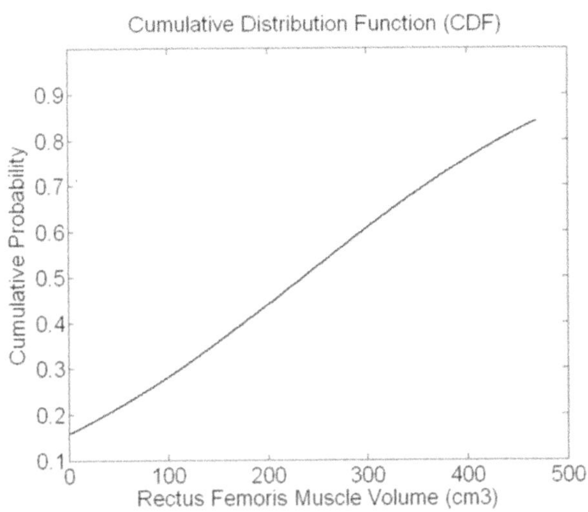

Figure 2.9. *Cumulative distribution function of the volume of the rectus femoris muscle*

Furthermore, PDF and CDF have many applications in risk analysis [SOH 02] or in structure analysis [PED 13, STI 13]. PDF and CDF are commonly used to model the random (aleatory) uncertainty.

2.2.1.2. *Mass function*

Classical uncertainty representation formalisms such as PDF and CDF showed their limitations for the uncertain and incomplete data modeling. Thus, other uncertainty representation formalisms and structures such as the mass function, plausibility function and possibility function could be used in this case.

The mass function is a structure of belief theory. The belief theory or the theory of evidence has been known as the Dempster–Shafer theory. This theory is a general framework for statistical inference as well as for uncertainty reasoning. Furthermore, structures of Dempster–Shafer theory such as mass functions and combination rules provide basic components for a fusion approach from different heterogeneous data sources. Let Ω = {C1, C2,..., Cn} denote the frame of discernment which is defined as a finite set of possible target values of a variable *X*. The mass function of each subset A of Ω is defined as follows:

$$m: 2^{\Omega} \rightarrow [0,1] \sum_{A \subseteq \Omega} m(A) = 1 \quad [2.8]$$

Each subset A with $m(A) > 0$ is called a focal set. The mass function *m* represents a body of evidence, ranging from 1 (perfect knowledge) to 0 (complete ignorance), relative to the value of *X* or a state of belief induced from this body of evidence. From mathematical point of view, *m* can be considered as a generated probabilistic distribution (i.e. mass distributes on the 2^{Ω} instead of Ω).

The construction of mass function is a challenging issue for the application of belief theory. Some approaches exist such as calculation of the mass function based on previous observations [DEN 06] or based on expert judgment [BRY 99]. We proposed a rule-based approach, which is simple and intuitive, to build mass functions without significant effort of experimentations. Basic belief assignments of each object $\left(m_{k,i}(C_k), m_{k,i}(\overline{C_k}), m_{k,i}(\Omega)\right)$ in the frame of discernment Ω are computed using a heuristic-based fuzzy rule illustrated in Figure 2.10. The min and max threshold values are parameterized according to the selection of Prop (Prop is one of the extracted useful features from temporal data).

$$
\begin{aligned}
& m_{k,i}(C_k) : f_{Prop\in[minThreshold,maxThreshold]}(Prop) \to [0,1] \\
& m_{k,i}(\overline{C_k}) : f_{Prop<minThreshold}(Prop) \to [1,0] \\
& m_{k,i}(\Omega) : f_{Prop<minThreshold}(Prop) \to [0,1] \\
& f_{Prop\in[minThreshold,maxThreshold]}(Prop) \to [1,0] \\
& \forall k \in \{1,\ldots,n\} \forall i \in \{1,\ldots,m\}
\end{aligned}
\qquad [2.9]
$$

where n is the number of objects (classes) of the frame of discernment Ω and m is the number of useful extracted features.

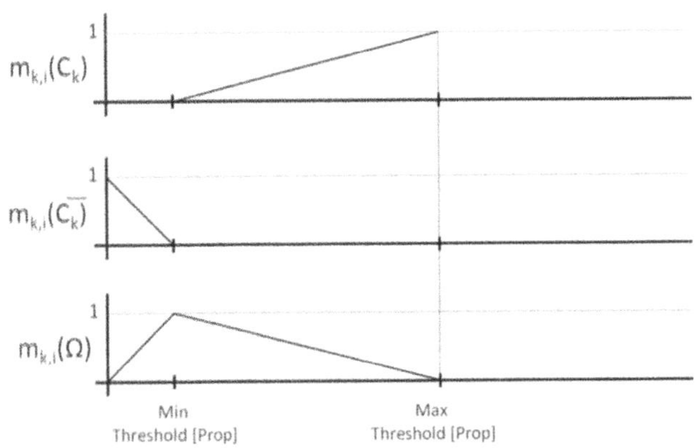

Figure 2.10. *Heuristic-based fuzzy rule for basic belief assignment*

EXAMPLE 2.6.– An example of basic belief assignments (mass function) of two classes of children with cerebral palsy (diplegia and hemiplegia) according to kinetics data is provided in Table 2.7.

Diplegia (C_1)			Hemiplegia (C_2)		
$m_{1,1}(C_1)$	$m_{1,1}(\overline{C_1})$	$m_{1,1}(\Omega)$	$m_{1,2}(C_2)$	$m_{1,2}(\overline{C_2})$	$m_{1,2}(\Omega)$
0.6087	0	0.3913	0.3108	0	0.6892
0.5362	0	0.4638	0.4054	0	0.5946
1.0000	0	0	1.0000	0	0
0.6280	0	0.3720	0.6552	0	0.3448
1.0000	0	0	1.0000	0	0
0.0145	0	0.9855	0.5811	0	0.4189
0.0362	0	0.9638	0.8514	0	0.1486
0.0435	0	0.9565	0.5270	0	0.4730
0.0870	0	0.9130	0.5946	0	0.4054

Table 2.7. *An example of basic belief assignments*

2.2.1.3. *Probability-box*

The probability-box (p-box) is a probability structure which simultaneously represents the random uncertainty and the epistemic uncertainty. The p-boxes [FER 03] approach was introduced recently with real potential applications such as the reliability analysis of polynomial systems [CRE 12], the evaluation of probabilistic sewer flooding [SUN 12] and the cost uncertainty analysis [MEH 12]. The p-boxes structures deal with non-parametric and parametric p-boxes with known sample distribution [FER 06]. Recently, the theoretical aspect of the p-boxes has been improved with new structures such as generalized p-boxes [DES 08] and Bayesian p-boxes with multiple random quantities and dependent parameters [MON 09]. The p-box of an observable random continuous parameter of interest that has the specified distribution is a parametric p-box consisting of four distributions $(D_i\left(\mu_i^l,\sqrt{\sigma_i^l}\right),D_i\left(\mu_i^l,\sqrt{\sigma_i^u}\right),D_i\left(\mu_i^u,\sqrt{\sigma_i^l}\right),D_i\left(\mu_i^u,\sqrt{\sigma_i^u}\right))$, where $\left[\mu_i^l,\mu_i^u\right]$ and $\left[\sigma_i^l,\sigma_i^u\right]$ are the mean and standard deviation intervals. A graphical representation of a p-box with a normal distribution assumption is shown in Figure 2.11. Note that a p-box is used to simultaneously express incertitude (epistemic uncertainty), which is represented by the breadth between the left and right edges of the p-boxes, and variability (random uncertainty), which is represented by the overall slant of the p-boxes.

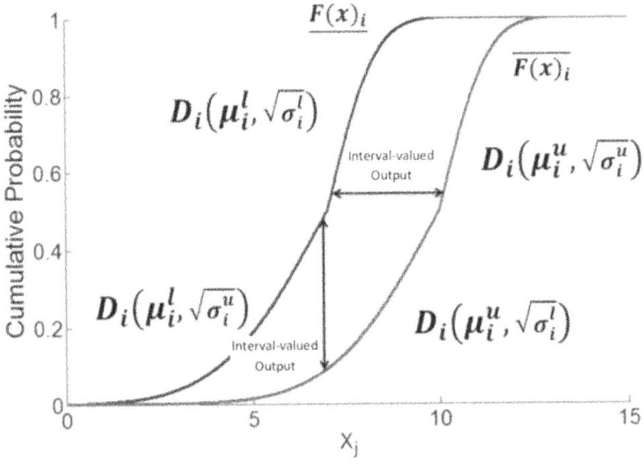

Figure 2.11. *Graphical representation of a p-box with normal distribution assumptions*

2.2.1.4. *Knowledge-based fusion probability-box*

The p-box needs to be improved in the case of data extracted from sources. We proposed recently new knowledge-based fusion p-boxes to model the random and epistemic uncertainties of biomechanics data from multiple sources or from new updated ones. These new structures integrate and aggregate the expert's judgments regarding the accuracy level of the experimental measuring protocol into the uncertainty modeling, leading to the improvement of the accuracy of the uncertainty representation model.

To construct the knowledge-based fusion p-boxes, a knowledge-based p-box is defined based on the p-box structure. Given a 5-druple value set $[\mu_i,\sigma_i,D_i,U_i^r,U_i^e]$ calculated from each source S_i where $\mu_i \in \left[\mu_i^l,\mu_i^u\right]$ is the interval of mean range of values of X_j, $\sigma_i \in \left[\sigma_i^l,\sigma_i^u\right]$ is the interval of standard deviation range of values of X_j and D_i is the known specified distribution of X_j. In particular, $U_i^r \in \{0,\ldots,1\}$ is defined as a random uncertainty coefficient of X_j, and $U_i^e \in \{0,\ldots,1\}$ is defined as an epistemic uncertainty coefficient of X_j. Note also that μ_i^l and σ_i^l can be equal to μ_i^u and σ_i^u, respectively, for some cases in which only one value for each mean and standard deviation property is provided.

The knowledge-based p-box F_i^K of X_j having the specified distribution D_i from source S_i is a parametric p-box consisting of four D_i distributions $(D_i\left(\mu_i^l,\sqrt{\sigma_i^l}\right),D_i\left(\mu_i^l,\sqrt{\sigma_i^u}\right),D_i\left(\mu_i^u,\sqrt{\sigma_i^l}\right),D_i\left(\mu_i^u,\sqrt{\sigma_i^u}\right))$, where the mean $\left[\mu_i^l,\mu_i^u\right]$ and standard deviation $\left[\sigma_i^l,\sigma_i^u\right]$ intervals are computed as follows:

$$\mu_i^l = \left|\mu_i - \gamma^\mu \times U_i^r\right| \quad [2.10]$$

$$\mu_i^u = \left|\mu_i + \gamma^\mu \times U_i^r\right| \quad [2.11]$$

$$\sigma_i^l = \left|\sigma_i - \gamma^\sigma \times U_i^e\right| \quad [2.12]$$

$$\sigma_i^u = \left|\sigma_i + \gamma^\sigma \times U_i^e\right| \quad [2.13]$$

where γ^μ and γ^σ are the scaling constants calculated as follows:

$$\gamma^\mu = \begin{cases} quotient\left(\dfrac{\mu^u}{10}\right), \ \mu^u \geq 10 \\ quotient\left(\dfrac{\mu^u}{10}\right) + 1, \text{ otherwise.} \end{cases} \quad [2.14]$$

and

$$\gamma^\sigma = \begin{cases} quotient\left(\dfrac{\sigma^u}{10}\right), \ \sigma^u \geq 10 \\ quotient\left(\dfrac{\sigma^u}{10}\right) + 1, \text{ otherwise.} \end{cases} \quad [2.15]$$

The calculation of the random uncertainty coefficient ($U_i^r \in \{0,\ldots,1\}$) of X_j from data source $S_i \in \{S_1, S_2, \ldots, S_k\}$ is based on a variability-based approach. The random uncertainty coefficient is calculated as follows:

$$\begin{aligned} &U_i^r = \frac{w_i}{w} \ w_i = x_i^u - x_i^l \ w = x^u - x^l \\ &x^l = Min\left\{\mu_i^l - \sigma_i^u\right\} \forall i \\ &x^u = Max\left\{\mu_i^u + \sigma_i^u\right\} \forall i \end{aligned} \quad [2.16]$$

The lower and upper probability-bound non-decreasing functions $\left(\underline{F(x)}_i \leq \overline{F(x)}_i\right)$ of F_i^K are formulated by using the following mathematical formulas:

$$\underline{F(x)}_i = CDF\left(\mu_i^l,(\sigma_i^u)^2\right)_1^{\mu_i^l\times\frac{1}{10^{-step}}} \cap CDF\left(\mu_i^l,(\sigma_i^l)^2\right)_{\mu_i^l\times\frac{1}{10^{-step}}}^{x^u\times 10^{step}} \quad [2.17]$$

$$\overline{F(x)}_i = CDF\left(\mu_i^u,(\sigma_i^l)^2\right)_1^{\mu_i^u\times\frac{1}{10^{-step}}} \cap CDF\left(\mu_i^u,(\sigma_i^u)^2\right)_{\mu_i^u\times\frac{1}{10^{-step}}}^{x^u\times 10^{step}} \quad [2.18]$$

where 10^{-step} is the fractional coefficient used to discretize the lower and upper bound set of X_j from all k data sources, and CDF means the cumulative distribution function. If the fractional coefficient of 0.01 is used, then step =2 (10^2).

The knowledge-based fusion p-box F_F^K of X_j from all k data sources is a combination of different knowledge-based p-boxes of each source S_i. The lower and upper probability-bound non-decreasing functions $\left(\underline{F(x)}_F \leq \overline{F(x)}_F\right)$ of F_F^K are formulated by using the following mathematical formulas:

$$\underline{F(x)}_F = Min\left\{\underline{F(x)}_1, \underline{F(x)}_2, \ldots, \underline{F(x)}_k\right\} \quad [2.19]$$

$$\overline{F(x)}_F = Max\left\{\overline{F(x)}_1, \overline{F(x)}_2, \ldots, \overline{F(x)}_k\right\} \quad [2.20]$$

Graphical illustration of a knowledge-based fusion p-box is shown in Figure 2.12.

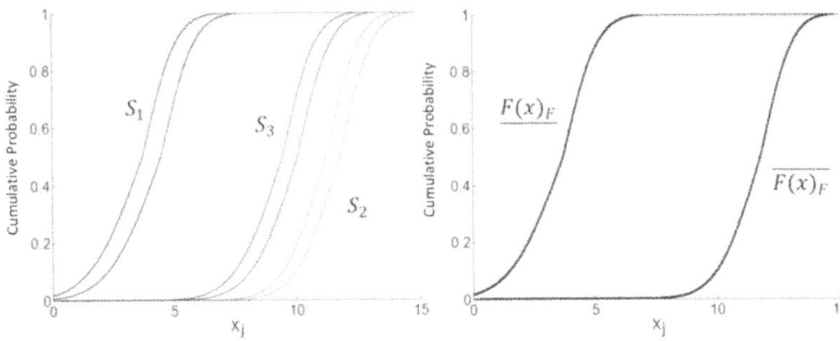

Figure 2.12. *Graphical representation of a knowledge-based fusion p-box with normal distribution assumptions*

2.2.2. *Uncertainty modeling*

2.2.2.1. *Biomechanical multiphysics model*

The biomechanical multiphysics model is a powerful tool for analyzing the mechanical functions of living systems such as the human musculoskeletal system. The development of a biomechanical multiphysics model requires advanced knowledge of human anatomy and physiology, and especially the pathophysiological process that occurs in the case of musculoskeletal diseases. A biomechanical multiphysics model is a mathematical representation of a simplified system of interest (e.g. the human musculoskeletal system). This representation connects the observable input data into the target output data through physical or biological laws. From a mathematical point of view, variations of input data play an essential role in the range of values of target output data. Thus, the uncertainty of input data strongly influences the output results. Consequently, this uncertainty needs to be modeled in order to control and efficiently manage its effect. In this section, we present a muscle rheological model, which is commonly used in the musculoskeletal modeling community. Then in the next section, the uncertainty of input data is modeled.

A muscle rheological model, such as the Hill-type model, aims to estimate the human *in vivo* muscle forces during motion. This model is one of the most used muscle rheological models to estimate the muscle forces through the input–output relationship (e.g. force–length or force–velocity relationships) of the muscle of interest. Consequently, the uncertainty of the input data (e.g. physiological cross-sectional area, optimal muscle-fiber length) has an impact on the output results (e.g. muscle tensile force). The simplest Hill-based muscle model consists of only one contractile element (CE). For this simplest Hill-based muscle model, the estimated muscle tensile force is not based on its length or velocity behavior. There are four input parameters: $a_M(t_i)$ which is the muscle activation as a function of the simulation time step $t_i(s)$; $pCSA_M(cm^2)$ which is the muscle physiological cross-sectional area; $\gamma_M(\mathrm{N}/cm^2)$ which is a muscle scaling factor; and F_M^0 which is the peak isometric muscle force. The output result is the tendon force F^T. The constitutive equations governing the input–output muscle behavior law are expressed as follows:

$$F^T = F^{CE} = a_M(t_i) \times F_M^0 \quad [2.21]$$

$$F_M^0 = pCSA_M \times \gamma_M \quad [2.22]$$

2.2.2.2. *Data uncertainty of muscle morphological properties*

In this section, we show how to use the knowledge-based fusion p-boxes to model the uncertainty of the physiological cross-sectional area parameter of the skeletal muscle. First, we collected all possible literature-based values of the physiological cross-sectional area $pCSA_M$ of the rectus femoris muscle from reliable sources such as ScienceDirect and PubMed. Related data are given in Table 2.8.

Data Source	$pCSA_M$ (cm²) $(\mu_i \pm \sigma_i)$
S_1 [WAR 09]	13.5 ± 5
S_2 [LIM 12]	8.3 ± 2
S_3 [HOR 07]	28.9
S_4 [NAR 92]	66.2 ± 12.1
S_5 [ARN 10]	13.9

Table 2.8. *Physiological cross-sectional area of the rectus femoris muscle from five separate data sources*

The lower and upper bounds of the $pCSA_M$ are calculated and shown in Figure 2.13. The total range of values of the rectus femoris $pCSA_M$ is [6.2, 78.3] (cm²).

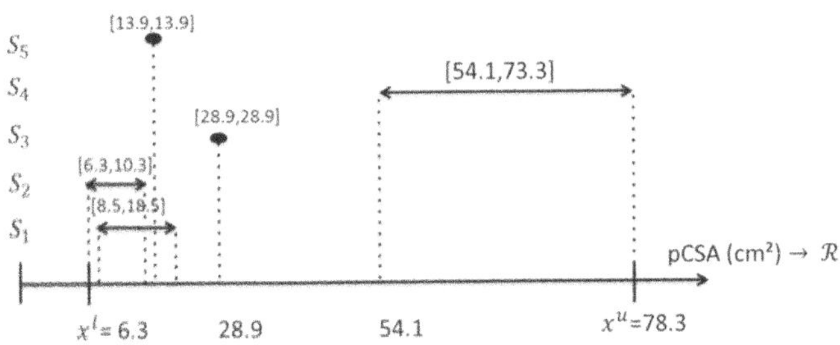

Figure 2.13. *Graphical representation of lower and upper data bounds*

The random uncertainty coefficients of the physiological cross-sectional area of the rectus femoris muscle are presented in Table 2.9. $w_{pCSA} = 78.3 - 6.3 = 72$ and $w_\gamma = 61 - 25 = 36$.

Two biomechanical experts (experts, respectively, A and B), having in 20 and 7 years approximately of qualifying work experience in the biomechanics domain, especially in musculoskeletal modeling, had participated in the computation of the epistemic uncertainty coefficients U_i^e for the $pCSA_M$ property. The rating scores of the two experts are shown in Table 2.9. Thus, the five-druple value sets with normal distribution assumption of the $pCSA_M$ are presented in Table 2.10. Finally, the knowledge-based fusion p-boxes of the physiological cross-sectional area of the rectus femoris muscle from five separate data sources are shown in Figures 2.14 and 2.15.

Parameter	U_i^r				
	S_1	S_2	S_3	S_4	S_5
$pCSA_M$	0.14	0.06	0	0.27	0

Table 2.9. *Random uncertainty coefficients of the physiological cross-sectional area of rectus femoris muscle from five separate data sources*

Parameter	U_i^e				
	S_1	S_2	S_3	S_4	S_5
$pCSA_M$	0.465	0.4	0.2	0.38	0.1

Table 2.10. *Epistemic uncertainty coefficients of the physiological cross-sectional area of the rectus femoris muscle from five separate data sources*

	$pCSA_M$		
	$[\mu_i, \sigma_i, D_i, U_i^r, U_i^e]$	$[\mu_i^l, \mu_i^u]$	$[\sigma_i^l, \sigma_i^u]$
S_1	$[13.5, 5, N, 0.14, 0.465]$	$[13.36, 13.64]$	$[4.54, 5.47]$
S_2	$[8.3, 2, N, 0.06, 0.4]$	$[8.24, 8.36]$	$[1.6, 2.4]$
S_3	$[28.9, 0, N, 0, 0.2]$	$[28.9, 28.9]$	$[0.2, 0.2]$
S_4	$[66.2, 12.1, N, 0.27, 0.38]$	$[64.58, 66.36]$	$[11.7, 12.48]$
S_5	$[13.9, 0, N, 0, 0.1]$	$[13.9, 13.9]$	$[0.1, 0.1]$

Table 2.11. *Five-druple value sets of the $pCSA_M$ (N means the normal distribution)*

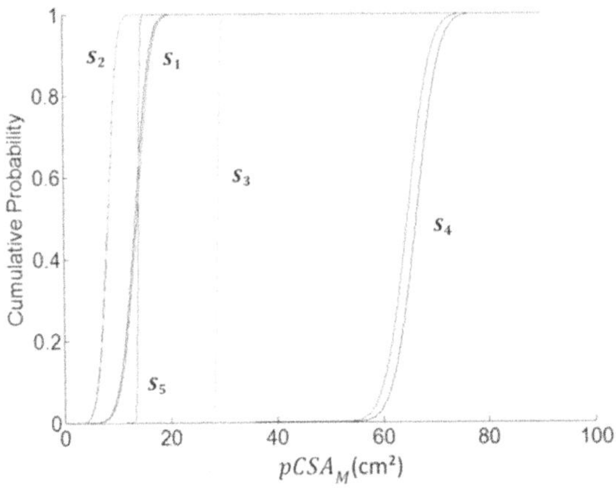

Figure 2.14. *Graphical representation of separate knowledge-based p-boxes of the physiological cross-sectional area from five separate data sources*

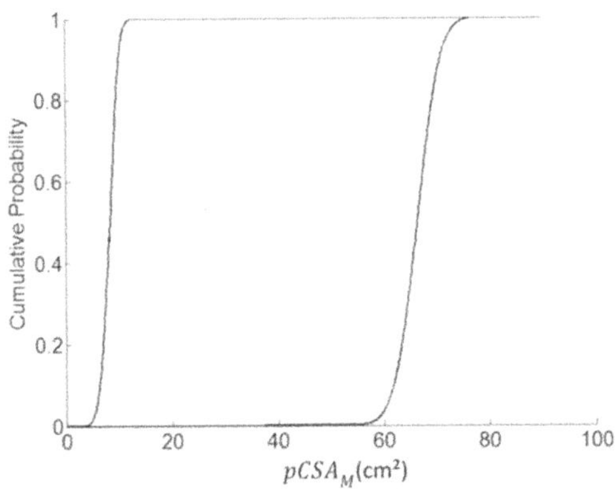

Figure 2.15. *Graphical representation of the knowledge-based fusion p-boxes of the physiological cross-sectional area from five separate data sources*

2.3. Biomechanical data uncertainty propagation

2.3.1. *Forward and backward uncertainty propagation*

The aim of the uncertainty propagation is to study the effect of input parameter uncertainties through a mathematical equation, leading to the quantification of the output response uncertainties. The input variables of our system could be either independent or dependent parameters.

Forward uncertainty propagation relates to the quantitative analysis of the impact of input uncertainty space on the corresponding output results. Backward uncertainty propagation deals with the reverse direction starting from the distribution of the output function to identify the distribution of input data uncertainty space as illustrated in Figure 2.16.

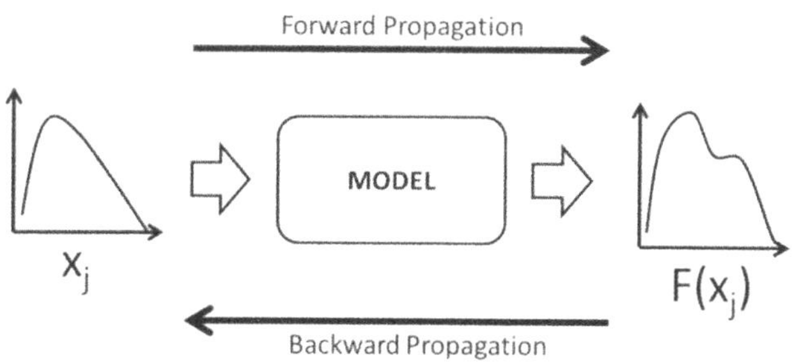

Figure 2.16. *Overview of uncertainty propagation approaches*

2.3.2. *Independent and dependent parameters*

Independent parameters are those parameters that have no relationship with other parameters. From a biomechanical point of view, the relationship relates to an anatomical or functional relationship, measuring relationship or statistical relationship. Inversely, dependent parameters have one or more relationships with other parameters. In the biomechanics field of study, dependent parameters are common. For example, the muscle length parameter has anatomical and functional relationships with the muscle volume parameter.

2.3.3. *Monte Carlo simulation*

The Monte Carlo method consists of the random selection of input variable samples $(X_j \in \mathcal{R}, j \in \{1,\ldots,p\})$ and the computation of a sequence of random output response $R = f(X_1, X_2, \ldots, X_p)$ based on the input variable samples. Then, a CDF of the computed sequence is formulated as shown in Figure 2.17. Monte Carlo simulation is commonly used in the case of uncertainty propagation of independent parameters. It is important to note that the random selection is essential for the Monte Carlo simulation. Thus, the random process needs a powerful mathematical function to achieve an effective random outcome.

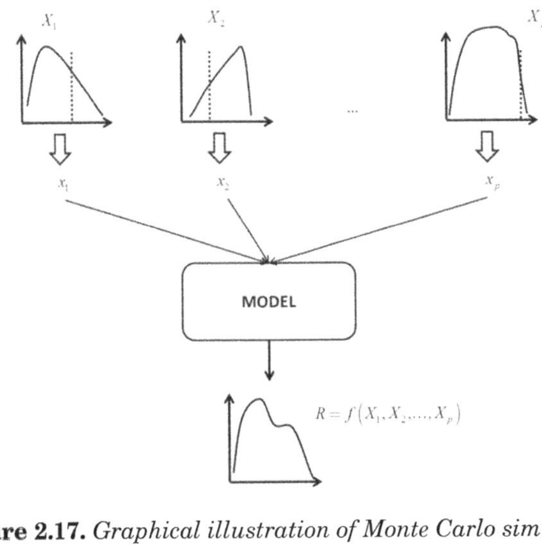

Figure 2.17. *Graphical illustration of Monte Carlo simulation*

2.3.4. *Copula-based Monte Carlo simulation*

For the case of uncertainty propagation of dependent parameters, the coupling between copula structure and Monte Carlo simulation could be used as a potential solution. A copula is a particular joint distribution function describing the dependence between random parameters. The basic idea of a copula is based on Sklar's theorem [SKA 59, SCA 89, NEL 06].

Sklar's theorem: let $X_j \in \mathcal{R}, j \in \{1,\ldots,p\}$ be random variables with joint distribution H and marginal distribution functions $\{G_1(x_1), G_2(x_2), \ldots, G_p(x_p)\}$. Then, there exists a copula C controlling the joint distribution such that:

$$C:[0,1]^2 \rightarrow [0,1] \quad [2.23]$$

$$\begin{aligned} &H(X_1, X_2, \ldots, X_p) = C(\{G_1(x_1), G_2(x_2), \ldots, G_p(x_p)\}) \\ &= \Pr(X_1 \le x_1, X_2 \le x_2, \ldots, X_p \le x_p) \end{aligned} \quad [2.24]$$

$$\forall x_i \in \Re$$

If $\{G_1(x_1), G_2(x_2), \ldots, G_p(x_p)\}$ are continuous functions, then C is unique. The constitutive equation of a two-parameter (X_1 and X_2) Gaussian copula is expressed as:

$$\begin{aligned} C_{Gaussian}(u, v; \rho) = \int_{-\infty}^{\Phi^{-1}(u)} \int_{-\infty}^{\Phi^{-1}(v)} \frac{1}{2\pi\sqrt{1-\rho^2}} exp \\ \left(-\frac{X_1^2 - 2\rho X_1 X_2 + X_2^2}{2(1-\rho^2)} \right) dX_1 dX_2 \end{aligned} \quad [2.25]$$

where u and v are cumulative (empirical) functions of standardized residuals and ρ is Pearson's linear correlation.

Thus, for the independent parameters, their lower and upper probability-bound functions ($\underline{F(x)}_F$ and $\overline{F(x)}_F$) of F_F^K are used to perform the random selection of input variable samples. For the dependent parameters, random selection of input variable samples is performed on the CFDs of their copulas. For this case, each random selection leads to all related samples. Then, the corresponding random output responses are computed using the mathematical governing function f. Finally, CDFs of the output responses are calculated as shown in Figure 2.18.

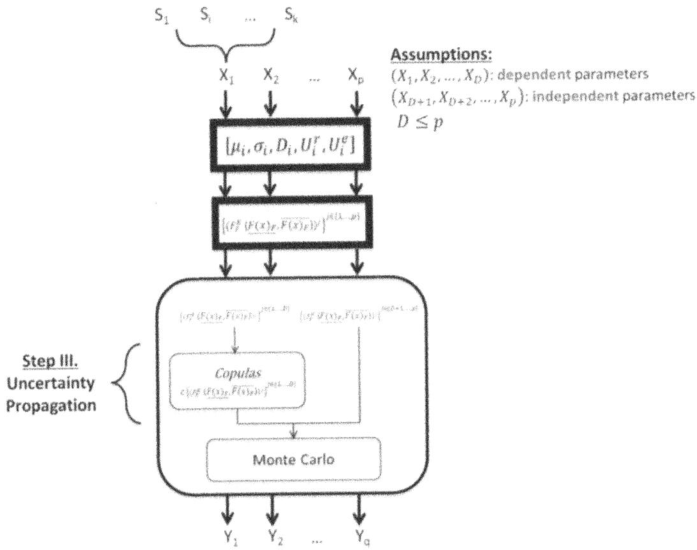

Figure 2.18. *Copula-based Monte Carlo simulation approach*

2.3.5. *Example of uncertainty propagation through a physical law*

In this section, the uncertainty propagation of the muscle's physiological cross-sectional area parameter is addressed through an example by using the forward uncertainty principle. The impact of the uncertainty of this parameter is measured through a physical law governed by a linear mathematical equation expressed by the equation $F_M^0 = pCSA_M \times \gamma_M$. The knowledge-based fusion p-boxes of the physiological cross-sectional area parameter are used. For the muscle scaling factor γ_M (N/ cm^2), the same methodology was applied to construct their knowledge-based fusion p-boxes as illustrated in Figures 2.19 and 2.20.

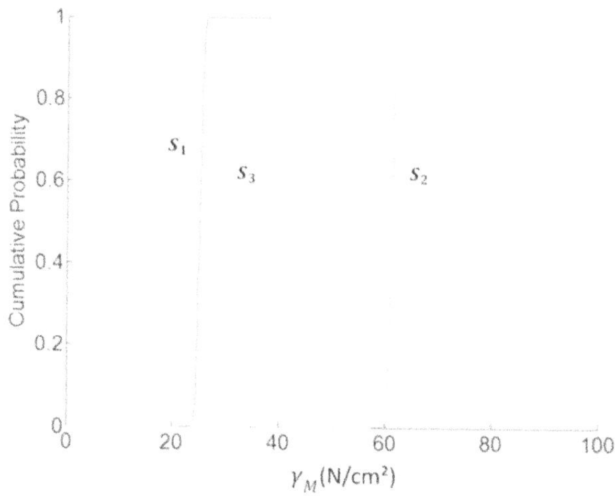

Figure 2.19. *Separate knowledge-based p-boxes of the muscle scaling factor parameters from three separate data sources*

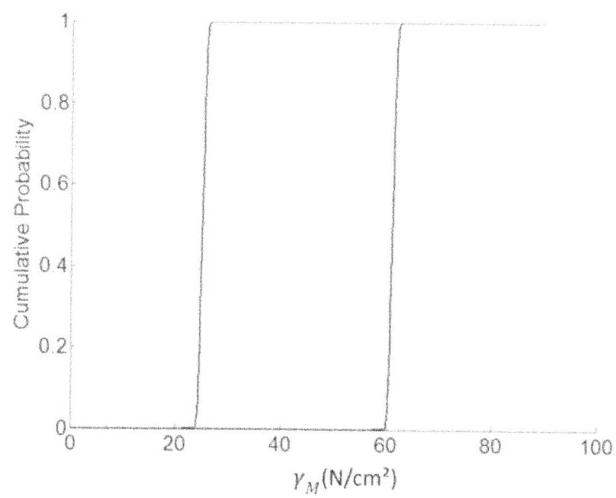

Figure 2.20. *Knowledge-based fusion p-boxes of the muscle scaling factor parameters from three separate data sources*

To maintain generality, a two-case simulation is presented. The first case relates to the assumption of an independent relationship between $pCSA_M$ and γ_M. The second case deals with the assumption of a dependent relationship between $pCSA_M$ and γ_M. For the independent case, the input uncertainties were propagated through the peak isometric muscle force computing model using a Monte Carlo sampling with 900 samples. After the propagation process, the distribution of the output response was estimated using an empirical CDF as shown in Figure 2.21. The empirical CDFs seem to be closely matched with the normal CDFs. Note that the Monte Carlo simulation was repeated 10 times. The results of the case of a dependent parameter assumption are shown in Figure 2.21. Related copulas of the physiological cross-sectional area and muscle tension factor parameters are shown in Figures 2.22 and 2.23.

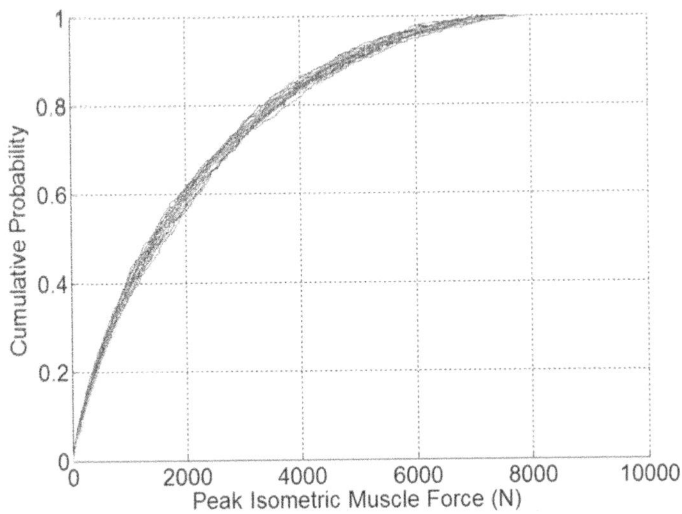

Figure 2.21. *Set of cumulative distribution functions generated by propagating the uncertainties of input independent data in the computation of the peak isometric muscle force: empirical CDFs from 10 repeated Monte Carlo runs*

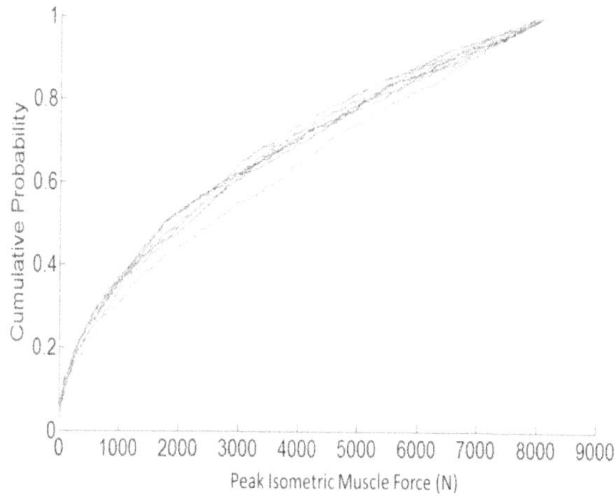

Figure 2.22. *Set of cumulative distribution functions generated by propagating the uncertainties of input-dependent data in the computation of the peak isometric muscle force: empirical CDFs from 10 repeated Monte Carlo runs*

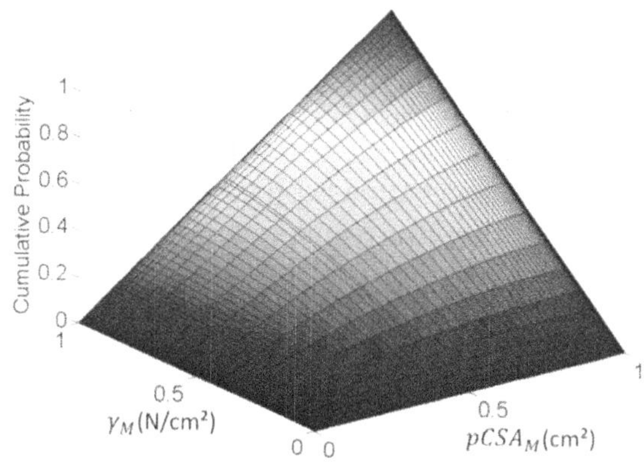

Figure 2.23. *CDFs of Gaussian copulas using marginal CDFs of the physiological cross-sectional area and muscle tension factor parameters*

2.4. Conclusions and perspectives

In this chapter, the modeling of biomechanical data uncertainties is addressed. Different types of biomechanical data are presented. The measuring chains of these data are analyzed to identify the possible uncertainty sources and the types of these data. A review of different uncertainty representation structures ranging from classical probability structures to new knowledge-based fusion p-boxes is provided. An example of the uncertainty modeling of the morphological properties of muscles is also presented. Finally, for uncertain propagation with different propagation approaches, a new copula-based Monte Carlo simulation is introduced. We believe that the data uncertainty modeling and uncertainty propagation play an essential role in the next generation of *in silico* multiphysics models used in a personalized medicine framework.

As far as perspectives are concerned, a roadmap is required to express how to determine the reduction strategy when the data uncertainties are modeled and mastered. New efforts will be investigated to study this challenging engineering subject.

2.5. Summary

– Biomechanical data types: physiological, morphological, mechanical and motion analysis parameters.

– Measuring chains: image-based and/or motion-based processing processes.

– Data uncertainty: random (aleatory) and epistemic uncertainties.

– Uncertainty sources: experimental protocol or numerical processing.

– Uncertainty representation structures: classical probability functions, new advanced structures such as mass function or knowledge-based fusion p-boxes.

– Uncertainty propagation: traditional Monte Carlo simulation for independent parameters and new copula-based Monte Carlo simulation for the dependent parameters.

– Challenging issue: a roadmap to express how to determine the reduction strategy when the data uncertainties are modeled and mastered.

2.6. Bibliography

[ARN 10] ARNOLD E.M., WARD S.R., LIEBER R.L., *et al.*, "A model of the lower limb for analysis of human movement", *Annals of Biomedical Engineering*, vol. 38, pp. 269–279, 2010.

[BAR 11] BARBER L., BARRETT R., LICHTWARK G., "Passive muscle mechanical properties of the medial gastrocnemius in young adults with spastic cerebral palsy", *Journal of Biomechanics*, vol. 44, pp. 2496–2500, 2011.

[BEN 06] BENSAMOUN S., RINGLEB S.I., LITTRELL L., *et al.*, "Determination of thigh muscle stiffness using magnetic resonance elastography", *Journal of Magnetic Resonance Imaging*, vol. 22, pp. 242–247, 2006.

[BRY 99] BRYSON N., MOBOLURIN A., "A process for generating quantitative belief functions", *European Journal of Operational Research*, vol. 115, pp. 624–633, 1999.

[CRE 12] CRESPO L.G., KENNY S.P., GIESY D.P., "Reliability analysis of polynomial systems subject to p-box uncertainties", *Mechanical Systems and Signal Processing*, 2012. Available at http://dx.doi.org/10.1016/j.ymssp.2012.08.012.

[DEN 06] DENOEUX T., "Constructing belief functions from sample data using multinomial confidence regions", *International Journal of Approximate Reasoning*, vol. 42, no. 3, pp. 228–252, 2006.

[DES 08] DESTERCKE S., DUBOIS D., CHOJNACKI E., "Unifying practical uncertainty representations – I: generalized p-boxes", *International Journal of Approximate Reasoning*, vol. 49, no. 3, pp. 649–663, 2008.

[FER 03] FERSON S., KREINOVICH V., GINZBURG L.R., *et al.*, Constructing probability boxes and dempster-shafer structures, Technical Report No. SAND2002-4015, Sandia National Laboratories, Albuquerque, New Mexico, pp. 1–143, 2003.

[FER 06] FERSON S., TUCKER W.T., "Sensitivity analysis using probability bounding", *Reliability Engineering and System Safety*, vol. 91, no. 10–11, pp. 1435–1442, 2006.

[HOR 07] HORSMAN K., KOOPMAN H.F.J.M., VAN DER HELM F.C.T., *et al.*, "Morphological muscle and joint parameters for musculoskeletal modelling of the lower extremity", *Clinical Biomechanics*, vol. 22, no. 2, pp. 239–247, 2007.

[LIM 12] LIMA K.M.E., DA MATTA T.T., DE OLIVEIRA L.F., "Reliability of the rectus femoris muscle cross-sectional area measurements by ultrasonography", *Clinical Physiology and Functional Imaging*, vol. 32, no. 3, pp. 221–226, 2012.

[MEH 12] MEHL C.H., "P-boxes for cost uncertainty analysis", *Mechanical Systems and Signal Processing*, 2012. Available at http://dx.doi.org/10.1016/j.ymssp.2012.03.014.

[MON 09] MONTGOMERY V.J., COOLEN F.P.A., HART A.D.M., "Bayesian probability boxes in risk assessment", *Journal of Statistical Theory and Practice*, vol. 3, pp. 69–83, 2009.

[NAR 92] NARICI M.V., LANDONI L., MINETTI A.E., "Assessment of human knee extensor muscles stress from in vivo physiological cross-sectional area and strength measurements", *European Journal of Applied Physiology and Occupational Physiology*, vol. 65, no. 5, pp. 438–444, 1992.

[NEL 06] NELSEN R.B., *An Introduction to Copulas*, Springer, 2006.

[PED 13] PEDRONI N., ZIOA E., FERRARIO E., *et al.*, "Hierarchical propagation of probabilistic and non-probabilistic uncertainty in the parameters of a risk model", *Computers & Structures*, 2013. Available at http://dx.doi.org/10.1016/j.compstruc.2013.02.003.

[POT 12] POTTER K., KIRBY R.M., XIU D., *et al.*, "Interactive visualization of probability and cumulative density functions", *International Journal for Uncertainty Quantification*, vol. 2, no. 4, pp. 397–412, 2012.

[RIN 07] RINGLEB S.I., BENSAMOUN S., CHEN Q., *et al.*, "Applications of magnetic resonance elastography to healthy and pathologic skeletal muscle", *J Magn Reson Imaging*, vol. 25, pp. 301–309, 2007.

[SCA 89] SCARSINI M., "Copulae of probability measures on product spaces", *Journal of Multivariate Analysis*, vol. 31, no. 2, pp. 201–219, 1989.

[SKA 59] SKLAR A., "Fonctions de répartition à n dimensions et leurs marges", *Publications de l'Institut de statistique de l'Université de Paris*, vol. 8, pp. 229–231, 1959.

[SOH 02] SOHRABI T.M., SHIRMOHAMMADI A., MONTAS H., "Uncertainty in nonpoint source pollution models and associated risks", *Environmental Forensics*, vol. 3, no. 2, pp. 179–189, 2002.

[STI 13] STIGSSON M., MUNIER R., "Orientation uncertainty goes bananas: an algorithm to visualise the uncertainty sample space on stereonets for oriented objects measured in boreholes", *Computers & Geosciences*, vol. 56, pp. 56–61, 2013.

[SUN 12] SUN S., FU G., DJORDJEVIC S., *et al.*, "Separating aleatory and epistemic uncertainties: probabilistic sewer flooding evaluation using probability box", *Journal of Hydrology*, vol. 420–421, pp. 360–372, 2012.

[WAR 09] WARD S.R., ENG C.M., SMALLWOOD L.H., *et al.*, "Are current measurements of lower extremity muscle architecture accurate?", *Clinical Orthopaedics and Related Research*, vol. 467, pp. 1074–1082, 2009.

3

Knowledge Modeling in Biomechanics of the Musculoskeletal System

Current clinical decision-making is commonly based on the patient data and personal knowledge of the clinician. Thus, the same patient with the same pathological state could be diagnosed and treated differently by different clinicians because each clinician has their own clinical knowledge base and the patient data could be different from one hospital or clinic to the next. This approach could lead to inefficient diagnosis and inappropriate treatment prescription in some clinical cases. To tackle this problem, clinical knowledge needs to be modeled objectively from real multimodal observable patient data to provide a generic knowledge base allowing evidence-based decision-making to be performed. This chapter addresses the conceptual and theoretical aspects of knowledge modeling in the biomechanics of the musculoskeletal system. First, the definition of the knowledge extraction and its modeling for specific applications in biomechanics are described. Second, essential steps ranging from knowledge representation (KR) and knowledge reasoning to conventional and advanced data mining process are presented. Finally, the conceptual and technical evolutions of evaluation-based systems ranging from expert systems, knowledge-based systems to healthcare systems of systems are illustrated and discussed.

3.1. Knowledge modeling in Biomechanics

3.1.1. *Introduction*

Knowledge modeling is a process of formalization of the domain knowledge to develop an interpretable model of knowledge, which could be processed and exploited by both human beings and computers. Knowledge modeling has been widely developed in the computer science field since the development of artificial intelligence in the 1950s [NIL 10]. Knowledge modeling in biomechanics of the musculoskeletal system is a novel field of study. It began in 2007 with the first published musculoskeletal ontology [DAO 07] used as a knowledge base for a clinical decision support (CDS) system [DAO 08]. The idea of the development of knowledge modeling in biomechanics of the musculoskeletal system arises from the fact that current biomechanical models could not be developed and applied for clinical applications without modeling assumptions and technical limitations. A knowledge model could be used as an alternative solution to provide better understanding of the mechanical behavior of the musculoskeletal system in interaction with internal and external conditions.

3.1.2. *Clinical benefits*

The use of knowledge modeling in biomechanics of the musculoskeletal system could lead to the following clinical benefits:

– Knowledge model (e.g. decision tree or belief decision tree) derived from supervised knowledge discovery (data mining) method could be used to assist the clinicians in their diagnosis process.

– Knowledge model (e.g. clustering model) derived from unsupervised knowledge discovery method could be used to propose an appropriate treatment according to the state of each patient. Moreover, this kind of model could be used to evaluate the effect or the quality of the treatment on the patient before and after its realization.

– All patient data could be used for the clinical decision-making.

– Clinical decision-making is based on a knowledge model derived statistically from a patient population.

– There is no modeling assumption for the development of the knowledge model.

3.2. Knowledge representation

3.2.1. *Web Ontology Language*

KR is a subresearch area of artificial intelligence aiming to represent knowledge in specific format such as symbols, semantic relationship and production rule (PR). KR allows new elements of knowledge to be induced from existing knowledge elements. There are many KR formalisms such as the Resource Description Framework (RDF) and RDF schema, Topic Maps, DARPA (Defense Advanced Research Projects Agency) Agent Markup Language (DAML), Ontology Inference Layer (OIL), Web Ontology Language (OWL) and PR. These formalisms have become standard frameworks to formalize the information and the knowledge about a domain of interest. In this chapter, we focus on two KR formalisms such as OWL and PR, which were already used in the biomechanics of the musculoskeletal system.

OWL is an XML-based KR language aiming to create and manipulate the ontology. Ontology is a structured specification of the concepts and semantic, intelligent relationships in a field of study [DAO 07]. OWL allows us to define the structure and relationship between information and knowledge through entity and property definitions. Moreover, it provides an advanced reasoning level of structured knowledge and information representation.

OWL is supported by the World Wide Web Consortium (W3C)[1]. There are many recommendations allowing us to define three variants of OWL such as OWL Lite, OWL

1 http://www.w3.org/TR/owl-guide/.

description logic (DL), and OWL Full. The difference of these variants is the level of expressiveness. OWL Full is the highest expressive language and the OWL Lite is the simplest expressive language. It is important to note that each variant is a syntactic extension of the precedent languages. For example, the OWL Full is a syntactic extension of the OWL DL and so on. To check the syntaxes and the species of the ontology, we could use the ontology validators such as the OWL validator[2] developed by the University of Manchester. The use of OWL to create linked information includes the definitions of the OWL header, OWL classes, subclasses and individuals, OWL properties, and OWL reasoning mechanisms.

EXAMPLE 3.1.– An example of the definition of a class "Bone" using OWL is provided as follows:

```
<owl:Classrdf:ID="Bone">

<rdfs:subClassOfrdf:resource="&BiologicalTissues"/>

<rdfs:labelxml:lang="en">Bone</rdfs:label>

<rdfs:labelxml:lang="fr">Os</rdfs:label>

<rdfs:labelxml:lang="vn">Xương</rdfs:label>

...

</owl:Class>
```

3.2.2. *Production rule*

PR is a KR formalism to organize the expert knowledge in form of a set of rules. A conventional PR has the form of **If***Condition* X_1 and/or *Condition* X_2 and/or.... *Condition* X_n **Then***Conclusion or Action Y*. Each rule could be fired using the real facts satisfying the set of conditions. A set of PRs is

2 http://owl.cs.manchester.ac.uk/validator/.

commonly used as a knowledge base of a traditional expert system such as MYCIN [SHO 76]. In particular, the combination of PRs and an inference engine allows the reasoning to be performed. In fact, a PR system commonly includes a set of PRs derived from expert knowledge, a working temporary memory and a reasoning mechanism using an inference engine.

EXAMPLE 3.2.– An example of the traditional set of PRs is provided as follows:

R1:

If Bone deformation = Yes and muscle paralysis = Yes

Then Pathology = Orthopedic disorder

R2:

If Pathology = Orthopedic disorder and muscle recoverable = Yes

Then Treatment strategy = Functional rehabilitation

3.3. Knowledge reasoning

Reasoning aims to perform a conclusion or decision or action by using the known facts or to determine what the facts needed for a specific conclusion or decision or action are. In this book, we focus only on the knowledge reasoning using PRs. Thus, there are three reasoning approaches such as forward chaining, backward chaining, and a hybrid combination of them. These reasoning approaches mimic human reasoning logics.

3.3.1. *Forward chaining*

Forward chaining is commonly used to discover the conclusion or decision or action, which could be derived from known facts as illustrated in Figure 3.1. During the process, an inference network is applied to fire the rule based on given facts satisfying the set of conditions governed by a set of Boolean operators (And, Or, Not).

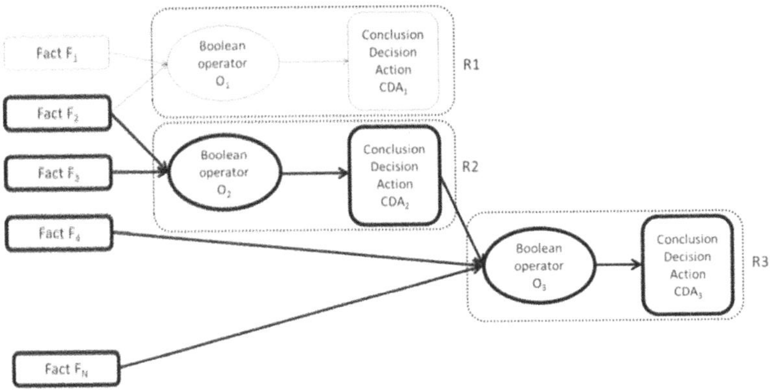

Figure 3.1. *Forward chaining: bold box represented given facts and fired conclusion / decision / action; bold connections represent the forward inference network*

3.3.2. *Backward chaining*

Backward chaining is commonly used to verify what the facts needed to perform a conclusion or decision or action are. For example, to determine if a clinical treatment is appropriate, we could use backward chaining to justify it by the facts through the backward inference network as illustrated in Figure 3.2.

3.4. Conventional and advanced knowledge discovery methods

3.4.1. *Knowledge discovery in databases*

The concept of data differs with those of knowledge and information. Data represent or describe an example or precise element or a fact of the world. Data could be verified using a reference in the real world. The detail of clinical observations of a patient is such a concept of data. Knowledge is the concept of know-how. Knowledge could be derived from deeper understanding gained through experiences or studies. It is important to note that experts are needed to acquire and formalize the knowledge. The interpretation of a pathological state based on patient data is such a kind of knowledge. Information is defined as a set of acquired data and knowledge to be shared among human beings or computer agents. The interaction between data, information, knowledge and decision is shown in Figure 3.3.

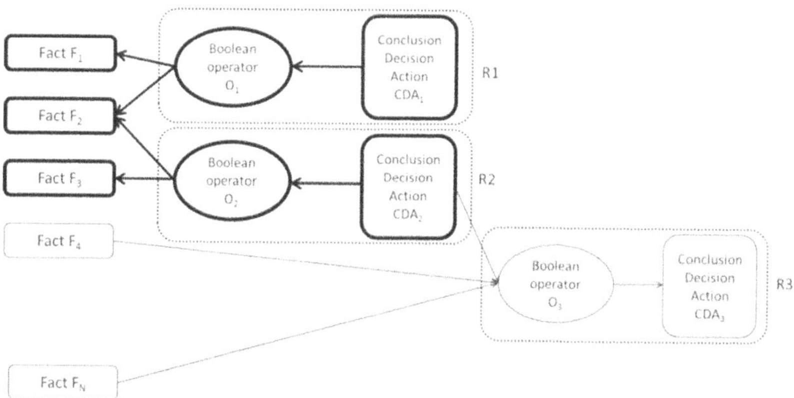

Figure 3.2. *Backward chaining: bold box represented given conclusion / decision / action and fired facts; bold connections represent the backward inference network*

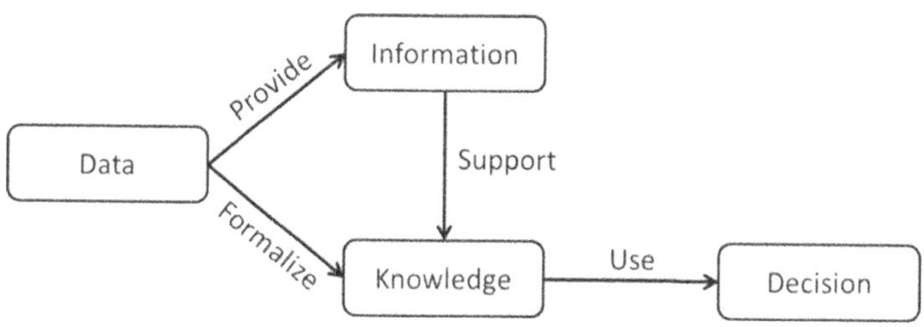

Figure 3.3. *Graphical illustration of the interaction between data, information and knowledge*

Knowledge discovery in databases is a process aimed at the extraction of knowledge from a huge quantity of data (i.e. Big data) using automatic or semi-automatic approaches [HAN 00, RAJ 13]. This analysis process is different from a statistical process. Statistical process allows us to verify *a priori* assumptions based on the observable data. Inversely, there are no *a priori* assumptions for the knowledge discovery analysis. This process aims to find out useful and pertinent knowledge from draw data. A knowledge management process is shown in Figure 3.4. From draw data, some target data are selected. Then, these data are processed to provide validated data, which is transformed into more useful data. Knowledge discovery is applied to develop a knowledge model, which could be visualized and used for decision-making purposes.

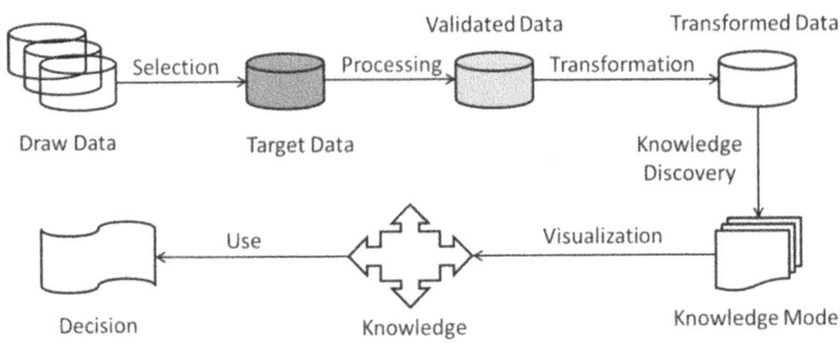

Figure 3.4. *Workflow of knowledge management process*

It is important to note that the knowledge discovery is only one step of the knowledge management process. Data mining is the analysis step in the process of knowledge discovery in databases. Data mining methods consist of two main approaches such as supervised classification and unsupervised classification (i.e. clustering). Supervised classification approach [HAN 01, ALO 12, CAR 13] uses given data and known labeled classes to develop a knowledge model such as a decision tree describing the statistical relationships between input data and output class. The unsupervised classification (i.e. clustering) method [GOW 84, FU 07] uses given data and unknown classes to develop a clustering model assigning a data set to a specific class. Recently, a new hybrid approach called semi-supervised classification has been introduced [MOU 10, YU 12]. This approach allows the classification model to be developed for given data, labeled and unlabeled classes.

A methodology of a knowledge discovery study in databases is shown in Figure 3.5. To develop a robust knowledge model, databases are divided into two sets such as a training set and testing set. A data mining algorithm is applied to the training set to develop the knowledge-based model. This model needs to be validated using the testing set. When validated, the model could be used for decision-making purposes with new patient data. It is important to note that the databases need to be representative and large enough to cover all possible patterns of the problem under investigation. Cross-validation could be an alternative solution in the case of small testing set.

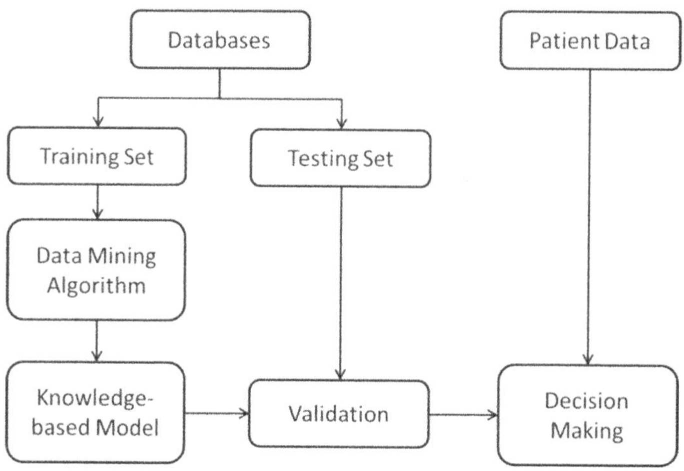

Figure 3.5. *Methodology of a knowledge discovery study*

From an applicative point of view, a classification model derived from knowledge discovery in databases could be used as a knowledge-based decision support tool. For example, clinician could use a knowledge model to assist his or her medical decision to perform a diagnosis or to prescribe an appropriate treatment planning. In the following sections, common supervised classification methods in biomechanics of the musculoskeletal system are described and discussed.

3.4.2. *Decision tree and belief decision tree*

3.4.2.1. *Classical decision tree*

Among supervised classification methods, a decision tree is the most self-explained decision support tool [ROK 08]. A decision tree is a tree-like graph or model of knowledge describing the statistical relationships between observable and measurable data and labeled classes as illustrated in Figure 3.6.

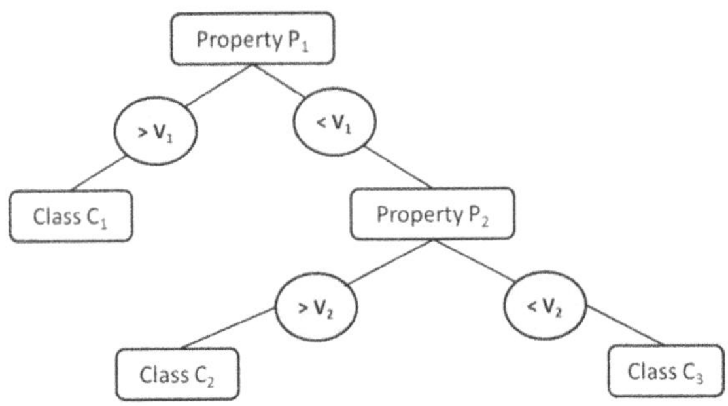

Figure 3.6. *Example of a decision tree with two measurable properties (P_1 and P_2), three labeled classes (C_1, C_2 and C_3) and two comparable values (V_1 and V_2)*

On the basis of decision tree, a set of PRs could be extracted and used for decision support purposes. Moreover, forward and backward chaining approaches could be applied to determine how to reach a class with given data or to justify what the data needed to lead a specific class are.

A decision tree could be created manually from the knowledge of a domain expert. This approach is very time consuming and becomes impractical when the decision tree is large. Thus, a decision tree is commonly created and tested using a database, which is divided into two sets: a training set and testing set. The training set is used to grow the decision tree while the testing set is used to evaluate the performance of the decision tree.

There are many existing algorithms (e.g. ID3, CHAID, C4.5, QUEST and random forest) for creating a decision tree. The most used algorithm is the C4.5 [QUI 93]. A decision tree has three principal elements: decision node, branch and class. To develop a decision tree from given data set, the following components needed to be taken into consideration: (1) attribute selection, (2) partitioning strategy, (3) stopping criterion and (4) leave structure.

Attribute selection: this requires a metrics to be used to select the best attribute to develop the decision tree from the root node. Entropy and gain information could be used as such a metrics. Let $\Theta = \{C_1, C_2, .., C_n\}$ be the set of labeled classes, A an attribute and its domain D(A) and T a training set. Each element in the training set T only belongs to only one labeled class. Gain information G could be computed using the following mathematical formulas:

$$G(T, A) = Info(T) - Info_A(T) \quad [3.1]$$

$$Info(T) = -\sum_{i=1}^{n} \frac{fre(C_i, T)}{|T|} \times \log_2 \frac{fre(C_i, T)}{|T|} \quad [3.2]$$

$$Info_A(T) = \sum_{v \in D(A)} \frac{|T_v|}{|T|} \times Info(T_v) \quad [3.3]$$

where $fre(C_i, T)$ is the number of elements in T having the labeled class C_i. T_v is a subset of elements in T having the value of attribute A equals to v. It is important to note that the best attribute is selected if its gain information $G(T, A)$ is maximal. Once selected, the growing process is iterative and repeated on each subset of the training set. One possible improvement of the gain information metrics is the use of gain ratio metrics GR, which is defined as follows:

$$GR(T, A) = \frac{G(T, A)}{SplitInfo(T, A)} \quad [3.4]$$

$$SplitInfo(T, A) = -Info_A(T) \quad [3.5]$$

Partitioning strategy: this is defined according to the data type. If the attribute is symbolic, the test of all possibilities is one of the partitioning strategies. If the attribute is numeric, its values could be discretized.

Stopping criterion: this needs to be defined to decide whether the tree could be grown or not. When all elements belong to their classes, the growing process of the decision tree could be stopped. The growing process could be stopping at the leaf level of the decision tree.

Leaf structure: this includes only one labeled class.

3.4.2.2. *Belief decision tree*

One of the advanced versions of the decision tree is the belief decision tree. This method is based on the classical decision tree and belief functions [ELO 01]. This method could be used to develop a decision tree in the framework of data uncertainty, for example the attributes are uncertain or missing and the class of each element is not a singleton. The growing process of a belief decision tree is the same as a traditional decision tree. It also requires the four following components to be taken into consideration: (1) attribute selection, (2) partitioning strategy, (3) stopping criterion and (4) leaf structure.

Attribute selection: this step aims to select the most appropriate attribute served as a decision node of the belief decision tree. There are two computing approaches: averaging and conjunctive approaches.

The *averaging approach* [ELO 01] is based on the pignistic probability entropy derived from the information entropy of the classical decision tree method. Given the mass function of each subset A of Ω is defined as $m: 2^{\Omega} \rightarrow [0,1]$ with $\sum_{A \subseteq \Omega} m(A) = 1$. The selection of the best attribute is performed using the following steps:

– Step 1: computing of the pignistic probability of each instance I_j of the training set T:

$$BetP^{\Theta}\{I_j\}(C_i) = \sum_{C_i \in C \in \Theta} \frac{1}{|C|} \times \frac{m^{\Theta}\{I_j\}(C)}{1 - m^{\Theta}\{I_j\}(\varnothing)}, \forall C_i \in \Theta \qquad [3.6]$$

– Step 2: computing of the average pignistic probability of each class on the set of elements S:

$$BetP^{\Theta}\{S\}(C_i) = \frac{1}{|S|} \sum_{I_j \in S} BetP^{\Theta}\{I_j\}(C_i) \qquad [3.7]$$

– Step 3: computing of the entropy of average pignistic probabilities in S:

$$Info(S) = -\sum_{i=1}^{n} BetP^{\Theta}\{S\}(C_i) \times \log_2 BetP^{\Theta}\{S\}(C_i) \qquad [3.8]$$

– Step 4: select the attribute A and create a subset S_v^A in which the attribute A has the value v.

– Step 5: computing of the average pignistic probability of each class on the set of elements S_v^A as follows:

$$BetP^{\Theta}\left\{S_v^A\right\}(\mathrm{C}_i), \mathrm{v} \in \mathrm{D(A)}, \mathrm{A} \in A_t, \mathrm{C}_i \in C \qquad [3.9]$$

– Step 6: computing of the information entropy of the attribute A as follows:

$$Info_A(S) = \sum_{v \in D(A)} \frac{\left|S_v^A\right|}{\left|S\right|} \times Info(\mathrm{S}_v^A) \qquad [3.10]$$

– Step 7: computing of the gain information provided by the attribute A in the set S as follows:

$$G(S, A) = Info(S) - Info_A(S) \qquad [3.11]$$

– Step 8: computing of the ratio gain of the following formulas:

$$\mathrm{RG}(S, A) = \frac{G(S, A)}{SplitInfo(S, A)} \qquad [3.12]$$

– Step 9: repeat for all attributes $\mathrm{A} \in A_t$ and select the best one maximizing the gain ratio RG.

The *conjunctive approach* [ELO 01] is based on the belief theory with the use of the distance between elements. The objective is to minimize the intragroup distance in maximizing the intergroup distance. The selection of the best attribute is performed using the following steps:

– Step 1: for each instance I_j of the training set T, compute the following value:

$$K\left\{I_j\right\}(C) = -\mathrm{lnq}^{\Theta}\left\{I_j\right\}(C), \forall C \subseteq \Theta \qquad [3.13]$$

– Step 2: for each value v of the attribute A, compute the following coefficient:

$$K\left\{S_v^A\right\}(C) = \sum_{I_j \in S_v^A} K\left\{J_j\right\}(C) \qquad [3.14]$$

– Step 3: compute the intragroup distance as follows:

$$SumD(S_v^A) = \frac{1}{|S_v^A|} \times \sum_{I_j \in S_v^A} \sum_{X \subseteq \Theta} (\mathrm{K}\{I_j\}(\mathrm{X}) - \frac{1}{|S_v^A|} \times \mathrm{K}\{S_v^A\}(\mathrm{X}))^2 \qquad [3.15]$$

– Step 4: for each attribute $\mathrm{A} \in A_t$, compute the following values:

$$SumD_A(S) = \sum_{v \in D(A)} \frac{|S_v^A|}{|S|} \times SumD(S_v^A) \qquad [3.16]$$

– Step 5: compute the intergroup distance as follows:

$$Diff(S, A) = SumD(S) - SumD_A(S) \qquad [3.17]$$

– Step 6: compute the difference ratio as follows:

$$DiffRatio(S, A) = \frac{Diff(S, A)}{SplitInfo(S, A)} \qquad [3.18]$$

– Step 7: repeat the computing process for each $\mathrm{A} \in A_t$ and select the best one maximizing the difference ratio.

Partitioning strategy: this is defined as the same strategy used in the case of classical decision tree.

Stopping criterion: the growing process of the belief decision tree is stopped when there is no attribute to verify or there is no better separation for the training set.

Leaf structure: this includes only one labeled class. Its mass function is computed using the following rules: (1) if the leaf includes only one element, its mass function equals that of this element; (2) if the leaf includes many elements, its mass function is computed as follows:

Averaging approach: $m^{\Theta}\{S\}(C) = \frac{1}{|S|} \sum_{I_j} m^{\Theta}\{I_j\}(C)$ [3.19]

Conjunctive approach: $m^{\Theta}\{S\} = \bigcap_{I_j \in S} m^{\Theta}\{I_j\}$ [3.20]

3.4.3. *Artificial neural network*

An artificial neural network is computing model inspired from the biological neural network. An artificial neural network is a flexible and adaptive model in which its structure could be changed to find patterns in a big data set [CHA 01]. A graphical representation of an artificial neural network denoted as a multilayer perceptron (MLP) [GAL 90] is shown in Figure 3.7. This type of network allows linear combination of input data using a connection-weighted principle. Another type of neural network is the radial basis function (RBF) [BUH 03] in which the distance between input data is taken into consideration.

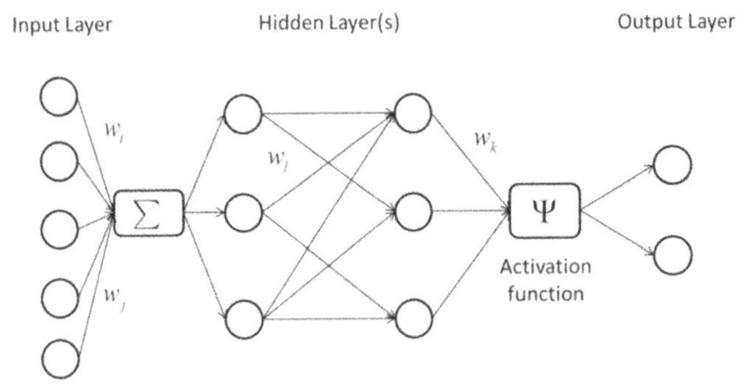

Figure 3.7. *Graphical representation of an artificial neural network*

3.4.4. *Support vector machine*

A support vector machine is the generalized case of the linear classifiers. Two main basic ideas are the maximal margin and the kernel function [COR 95, SUY 99]. The maximal margin is defined as the distance between the separate hyperplane and support vectors defined as the nearest data to the hyperplane. The kernel function allows the scalar product in large dimension space to be transformed into an evaluation of one simple function. This approach could be used to perform biclasses or multiclasses classification. Graphical illustration of a biclasses classification using a support vector machine is shown in Figure 3.8.

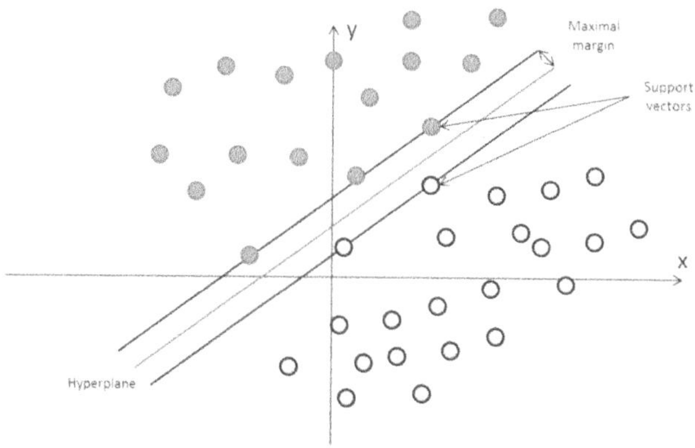

Figure 3.8. *Graphical representation of a biclasses classification using a support vector machine*

3.5. CDS system

A decision support system is a computer-aided program that could be used to assist clinicians in their decision-making. It is important to note that the aim of a CDS system is not to replace the clinicians but to assist them in performing a better diagnosis or to propose a more appropriate treatment prescription. Recently, a group of researchers proposed 10 grand challenges in CDS [SIT 08] as follows: (1) improve the human–computer interface, (2) disseminate the best practices in CDS design, development and implementation, (3) summarize patient-level information, (4) prioritize and filter recommendations to the user, (5) create an architecture for sharing executable CDS modules and services, (6) combine recommendations for patients with co-morbidities, (7) prioritize CDS content development and implementation, (8) create internet-accessible CDS repositories, (9) use free-text information to drive CDS and (10) nine large clinical databases to create new CDS. In this section, different generations of CDS systems are presented and discussed.

3.5.1. *Expert system*

An expert system is the first generation of the decision support system. The software architecture of a first-generation CDS system is illustrated in Figure 3.9. This system includes a knowledge base, an inference engine and a user interaction module. The knowledge base is commonly extracted directly from expert knowledge and stored in the form of a set of PR (IF ...AND...THEN ...). The inference engine consists of forward and backward chaining algorithms. The user interaction module provides only text-based dialogue between the expert system and the end users (e.g. clinicians).

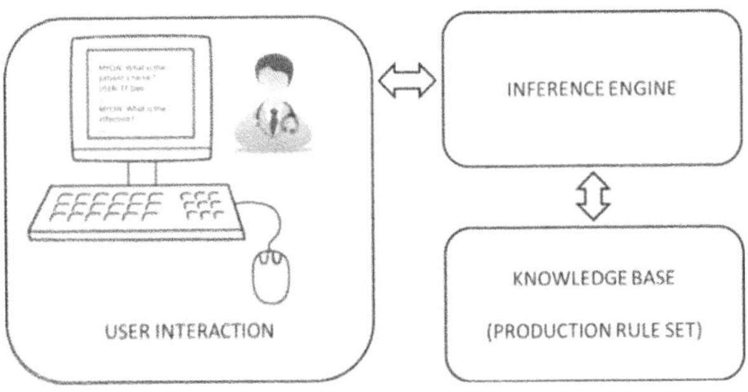

Figure 3.9. *Software architecture of a first-generation clinical decision support system*

MYCIN is one of the first expert systems. MYCIN was developed by Stanford University in the 1970s [SHO 76]. This expert system aims to identify the bacteria and recommend an antibiotics dosage according to the body weight of the patient. The knowledge base of the MYCIN system includes around

500 PRs. The accuracy of the MYCIN system is approximately 69% of correct diagnosed cases.

The first generation of the expert systems (e.g. MYCIN and CADUCEUS [BAN 86]) has a strong impact on research but these first expert systems were never used in routine practice. One of the main reasons relates to the fact that in the 1970s, the use of computer program in medical diagnosis created a debate about the responsibility of a wrong diagnosis performed by the computer. Another important reason deals with the limited software development technologies to create a user-friendly and helpful system for a clinician. For example, MYCIN provides only poor user interaction. The executing algorithm was very resource-demanding and time-consuming. Despite not being used in routine practice, the first generation of the expert systems was successful from a research point of view, giving the potential application of artificial intelligence theories, methods and techniques to the development of a CDS system.

3.5.2. *Knowledge-based system*

A knowledge-based system could be considered as a second-generation of a CDS system. Knowledge-based systems are currently being developed around the world to solve complex clinical problems using available knowledge and reasoning mechanisms. For this second-generation system, knowledge is commonly extracted from multiple sources (expert knowledge and databases) and represented by various KR formalisms such as classical and/or probabilistic PRs, frames and cases. In particular, artificial intelligence is the main core of such a system. By using traditional (e.g. a decision tree) and/or advanced data mining methods (e.g. a belief decision tree), useful knowledge could be extracted from a huge amount of data to provide evidence-based assets (e.g. a PR base). And then, statistical inference could be performed to provide predictive models for clinical purposes. The software architecture of a second-generation CDS system is illustrated in Figure 3.10.

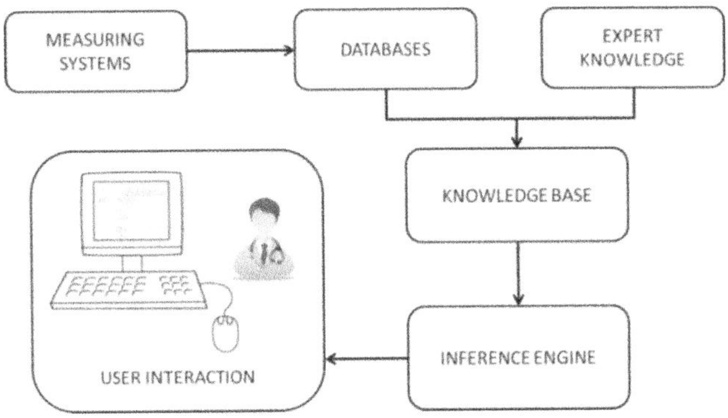

Figure 3.10. *Software architecture of a second-generation clinical decision support system*

3.5.3. *System of systems*

Knowledge-based systems have proven validity for the clinical and healthcare problems by using biomedical informatics approaches. However, this approach could detect only simple statistical relationships between data. More useful information or knowledge about anatomical or biological or physical causal relationships cannot be detected and extracted in a straightforward manner. Moreover, the accuracy of the predictive model is based on the generalization of the database under investigation. Recently, we proposed the integration of the knowledge-based system (i.e. a decision support system) with the multi-physical modeling one as a next-generation decision support system for performing personalized clinical recommendations. In short, a multi-physical model provided by a multi-physical modeling system is a personalized model describing the musculoskeletal structures (e.g. bone, muscle, tendon and ligament) and their mechanical behaviors. However, many modeling assumptions (such as muscle modeling, joint properties and functions) were performed. Consequently, a multi-physical model cannot integrate all accumulated knowledge (e.g. mechanical properties and behaviors) and experimental data due to their difficult implementations (e.g. appropriate formulation of mechanical behaviors, efficient algorithms and computing time). On the other hand, a knowledge-based model provided by a knowledge-based system is a predictive statistical model derived from one or multiple data sources. Thus, it can integrate all accumulated knowledge and data to improve the accuracy of the diagnosis, treatment and monitoring processes. Moreover, computational results of a multi-physical model could be used as input data of the knowledge-based model. In fact, these two modeling approaches are closely complementary. Thus, the integration of these approaches into a next-generation CDS system should be of great interest; benefiting their complementary advantages and limiting their weakness for performing efficient and personalized clinical recommendations. It is important to note that a next-generation decision support system needs to respond to the following requirements [MIT 11]: incorporating lessons from history, uniform vocabularies, integrative interfaces, contextualized decisions, personalized recommendations and adaptive solutions.

The notion of system of systems (SoS) has been recently introduced in the engineering system field [MAI 98]. From an engineering point of view, a system is defined as a group of functionally, physically and/or behaviorally interactive, independent, material or non-material components. An SoS is a set of useful systems integrated into a larger system to achieve a unique set of tasks [JAM 08]. Recently, a healthcare SoS was introduced to analyze and exploit the human brain as well as the orthopedic kinematic analyses using medical imaging techniques such as 2D X-ray fluoroscopy, ultrasound or magnetic resonance imaging [HAT 09]. In fact, an SoS approach could be an appropriate engineering approach to integrate the knowledge-based modeling system and the physics-based simulation system into an innovative decision support system for clinical purposes.

An example of the software architecture of a biomechanical SoS is shown in Figure 3.11. This CDS system consists of the following constituent systems:

– A data management system aiming to manage the multidimensional (morphological, mechanical, kinematic, kinetic and EMG) and multimodal (medical imaging techniques and 3D motion capture) data from different data acquisition sources. This system consists of data preprocessing and database modules.

– An ontology and information retrieval system dealing with a musculoskeletal ontology [DAO 07] served as common vocabularies used in our clinical DSS and as a knowledge kernel of information retrieval for web-based human musculoskeletal resources [DAO 13]. This system integrates an ontology module and a knowledge-based search engine module.

– A physics-based modeling system dealing with the modeling and simulation of the human musculoskeletal system in interaction with the external environment using mechanical engineering approaches [DAO 12]. This system integrates a multi-physics simulation module and a validation module.

– A knowledge-based system relating to the modeling of the musculoskeletal system using knowledge-based engineering approaches such as advanced data mining and artificial intelligence methods to perform statistical inference functions [DAO 11a, DAO 11b]. This system integrates an advanced statistical inference module and a validation module.

– An integration system aiming to aggregate knowledge from multi-physics simulation and knowledge-based modeling to provide evidence-based facts and knowledge for clinical decision-making. This system consists of a knowledge aggregation module and a decision-making module.

– A user interaction system aiming to manage the interaction between the results of our biomechanical SoS and the end users such as clinicians or biomedical researchers or biomedical engineers. This system consists of visualization and user online/offline interaction modules.

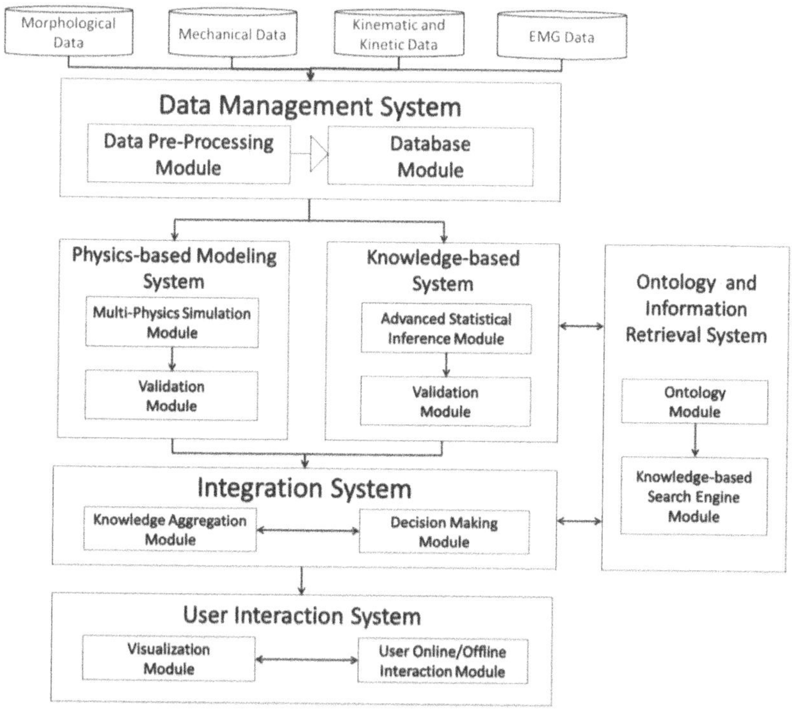

Figure 3.11. *Software architecture of the biomechanical system of systems*

3.6. Conclusions

In this chapter, theoretical concepts of knowledge modeling in the biomechanics of the musculoskeletal system are addressed. Different steps of a knowledge extraction process are presented and detailed. It is important to note that decision-making is always based on the knowledge of the pathologies under investigation. This knowledge could come from a personal (individual) base or from a more objective base. Thus, knowledge modeling of the investigated pathology will allow the diagnosis or treatment evaluation to be performed in an objective and quantitative manner. Furthermore, the use of a knowledge model needs to develop a CDS system. Among different software architectures, the SoS approach could be considered as a potential solution for the development of a next-generation CDS system.

3.7. Summary

– KR: OWL and PR.

– Knowledge reasoning: forward chaining and backward chaining.

– Data, information and knowledge: three basic elements of a knowledge extraction process.

– Clinical decision-making: diagnosis, treatment and prevention of a disease.

– CDS system: useful tool to assist the clinicians in their decision-making.

3.8. Bibliography

[ALO 12] ALONSO A.M., CASADO D., ROMO J., "Supervised classification for functional data: a weighted distance approach", *Computational Statistics & Data Analysis*, vol. 56, no.7, pp. 2334–2346, 2012.

[BAN 86] BANKS G., "Artificial intelligence in medical diagnosis: the INTERNIST/CADUCEUS approach", *Critical Reviews in Medical Informatics*, vol. 1, no. 1, pp. 23–54, 1986.

[BUH 03] BUHMANN M., *Radial Basis Functions: Theory and Implementations*, Cambridge University Press, 2003.

[CAR 13] CARRIZOSA E., MORALES D.R., "Supervised classification and mathematical optimization", *Computers & Operations Research*, vol. 40, no. 1, pp. 150–165, 2013.

[CHA 01] CHAU T., "A review of analytical techniques for gait data. Part 2: neural network and wavelet methods", *Gait and Posture*, vol. 13, pp. 102–120, 2001.

[COR 95] CORTES C., VAPNIK V.N., "Support-vector networks", *Machine Learning*, vol. 20, pp. 273–297, 1995.

[DAO 07] DAO T.T., MARIN F., HO BA THO M.C., "Ontology of the musculoskeletal system of lower limbs", *Proceedings of IEEE Engineering in Medicine and Biology Society*, pp. 386–389, 2007.

[DAO 08] DAO T.T., MARIN F., HO BA THO M.C., "Ontology-based computer-aided decision system: a new architecture and application concerning the musculoseletal system of the lower limbs", *IFMBE*, Springer, Berlin, Heidelberg, pp. 1540–1543, 2008.

[DAO 11a] DAO T.T., HO BA THO M.C., "Knowledge-based System for orthopedic pediatric disorders", *IFMBE Proceedings*, vol. 37, pp. 125–128, 2011.

[DAO 11b] DAO T.T., MARIN F., BENSAHEL H., *et al.*, "Computer-aided decision system applied to the clubfeet deformities", *Advances in Experimental Medicine and Biology*, vol. 696, no. 7, pp. 623–635, 2011.

[DAO 12] DAO T.T., MARIN F., POULETAUT P., *et al.*, "Estimation of accuracy of patient specific musculoskeletal modeling: case study on a post-polio residual paralysis subject", *Computer Method in Biomechanics and Biomedical Engineering*, vol. 15, no. 7, pp. 745–751, 2012.

[DAO 13] DAO T.T., HOANG T.N., TA X.H., *et al.*, "Knowledge-based personalized search engine for the web-based human musculoskeletal system resources (HMSR) in biomechanics", *Journal of Biomedical Informatics*, vol. 46, no. 1, pp. 160–173, 2013.

[ELO 01] ELOUEDI Z., MELLOULI K., SMETS P., "Belief decision trees: theoretical foundations", *International Journal of Approximate Reasoning*, vol. 28, pp. 91–124, 2001.

[FU 07] FU J., GAO H., FRANK J., "Unsupervised classification of single particles by cluster tracking in multi-dimensional space", *Journal of Structural Biology*, vol. 157, no. 1, pp. 226–239, 2007.

[GAL 90] GALLANT S.I., "Perception-based learning algorithms", *IEEE Transactions on Neural Networks*, vol. 2, pp. 179–191, 1990.

[GOW 84] GOWDA K.C., "A feature reduction and unsupervised classification algorithm for multispectral data", *Pattern Recognition*, vol. 17, no. 6, pp. 667–676, 1984.

[HAN 00] HAN J., KAMBER M., *Data Mining: Concepts and Techniques*, Morgan Kaufmann Publishers, 2000.

[HAN 01] HAND D.J., LI H.G., ADAMS N.M., "Supervised classification with structured class definitions", *Computational Statistics & Data Analysis*, vol. 36, no. 2, pp. 209–225, 2001.

[HAT 09] HATA Y., KOBASHI S., "Human health care system of systems", *IEEE Systems Journal*, vol. 3, no. 2, pp. 231–238, 2009.

[JAM 08] JAMSHIDI M., *Systems of Systems Engineering – Principles and Applications*, CRC/Taylor & Francis, London, 2008.

[MAI 98] MAIER M.W., "Architecting principles for system of systems", *Systems Engineering*, vol. 1, no. 4, pp. 267–284, 1998.

[MIT 11] MITCHELL J.A., GERDIN U., LINDBERG D.A.B., *et al.*, "50 years of informatics research on decision support: what's next", *Methods of Information in Medicine*, vol. 50, no. 6, pp. 525–535, 2011.

[MOU 10] MOUCHAWEH M.S., "Semi-supervised classification method for dynamic applications", *Fuzzy Sets and Systems*, vol. 161, no. 4, pp. 544–563, 2010.

[NIL 10] NILSSON N.J., *The Quest for Artificial Intelligence: A History of Ideas and Achievements*, Cambridge University Press, 2010.

[QUI 93] QUINLAN J.R., *C4.5 Programs for Machine Learning*, Morgan-Kaufmann Publishers, San Francisco, 1993.

[RAJ 13] RAJARAMAN A., LESKOVEC J., ULLMAN J., *Mining of Massive Datasets*, Cambridge University Press, 2013. Available at http://infolab.stanford.edu/~ullman/mmds/book.pdf.

[ROK 08] ROKACK L., MAIMON O., *Data Mining with Decision Trees: Theory and Applications*, World Scientific Publishing Company, 2008.

[SHO 76] SHORLIFFE E.H., *Computer Based Medical Consultations: MYCIN*, Elsevier, New York, 1976.

[SIT 08] SITTIG D.F., WRIGHT A., OSHEROFF J.A., *et al.*, "Grand challenges in clinical decision support", *Journal of Biomedical Informatics*, vol. 41, no. 2, pp. 387–392, 2008.

[SUY 99] SUYKENS J.A.K., VANDEWALLE J.P.L., "Least squares support vector machine classifiers", *Neural Processing Letters*, vol. 9, no. 3, pp. 293–300, 1999.

[YU 12] YU G., ZHANG G., DOMENICONI C., *et al.*, "Semi-supervised classification based on random subspace dimensionality reduction", *Pattern Recognition*, vol. 45, no. 3, pp. 1119–1135, 2012.

4

Clinical Applications of Biomechanical and Knowledge-based Models

In silico medicine is one of the challenging research areas in the field of biomechanics. This relates to the direct use of computer-based models and simulation in the diagnosis, treatment or prevention of a disease. In this chapter, clinical applications of biomechanical and knowledge-based models are expressed. The first application deals with the use of a patient specific musculoskeletal model to evaluate the effect of the orthosis on the functional behavior of a post-polio residual paralysis (PPRP) subject. The second application relates to the use of a computational ontology of the musculoskeletal system of the lower limbs in the interpretation of a cause–effect relationship of the rotational abnormalities. The third application expresses the development of predictive models of the pathologies (rotational abnormalities and clubfoot deformities) of the lower limbs.

4.1. Patient-specific musculoskeletal model: effect of the orthosis

4.1.1. *Introduction*

Orthopedic disorders such as children with cerebral palsy or PPRP affected the peripheral nervous and musculoskeletal systems leading to muscle spasticity or muscle paralysis and bony deformities [FAR 10, EUN 12]. The consequence of these disorders strongly influences the biomechanical functions (e.g. weight support or gait behaviors) of the human body of the involved patients

[WAK 07, DAO 12]. One of the most used clinical routine practices is the prescription of medical helping devices such as orthoses or crutches to recover normal motion behavior as well as to provide compensating aid for body support [FAR 10, LAM 05, GAR 06, GEN 10]. However, the impact of the orthosis on the biomechanical functions of the human body is still not well understood. Thus, inappropriate effects of the orthosis could make the handicap problem more serious for the patients. Furthermore, the current design of orthosis devices is based on laborious manual processes subject to human error (e.g. subjective measurements) and unfitted positions, not to mention being time consuming. All this can lead to bad quality clinical treatment and a high cost for medical activity [DEA 95, MAV 11]. Consequently, new specifications need to be recommended to improve the orthosis design process.

Orthoses such as ankle-foot orthosis (AFO) [CRU 09] or knee-ankle-foot orthosis (KAFO) [CUL 09] or hip-knee-ankle-foot orthosis (HKAFO) [FAR 10] are often designed to recover the biomechanical functions of the involved joints such as the hip, knee, ankle or a combination of the three. The choice the appropriate orthosis depends on the type as well as the pathological state of the patient. For example, a KAFO is commonly prescribed for a post-polio patient with paralyzed muscles on one side. Recently, the 3D motion capture technique has become a potential experimental technique to assess the performance of orthosis devices. Some research studies reported the improved locomotion patterns such as spatiotemporal parameters (gait speed and stride length), oxygen consumption, mechanical work and energy cost [LAM 05, CHE 05, DES 06, BLE 08]. However, these orthoses also showed intrinsic design limitations such as materials which were either too flexible or not flexible enough, inflexible and misadapted joints, inappropriate aids for gait kinematics and kinetics patterns as well as uncomfortable and unaesthetic patterns [LAM 05, BLE 08]. Furthermore, the pathophysiological state of the orthopedic patient could make the use of orthoses inefficient in some cases due to patient motion limitations. In fact, spatiotemporal or kinematics or kinetics data derived from 3D motion capture systems become qualitative and quantitative assessment properties of the consequence effect derived from the compromise between pathophysiological patient state and orthoses restrictions. However, the computing of these properties is based directly on the generic simplified biomechanical models provided with the motion capture systems such as VICON or visual 3D. It is well known that these generic biomechanical models do not accurately reflect the geometry of the patient involved. Consequently, a patient-specific biomechanical model would be of interest to provide more accurate data for the assessment of the orthosis effect.

The objective of this section is to present the quantification of the effect of a KAFO using a patient-specific biomechanical model derived from medical imaging and 3D motion capture techniques. On the basis of this evaluation, patient-specific KAFO orthosis design could be improved.

4.1.2. *Materials and methods*

4.1.2.1. *Data acquisitions*

CT scanner images were acquired on a PPRP patient (male, 26 years old, 170 cm height and 66 kg body mass) in a supine position. This patient was diagnosed and followed by an experienced radiologist and the orthopedic surgeon at the Polyclinique Saint Côme (Compiègne, France). Consequently, all parameters of the CT scan acquisition protocol were set up as clinical routine practice. A spiral-imaging scanner (GE Light Speed VCT 64) was used to perform our imaging acquisition protocol. Also note that the participated subject signed an informed consent agreement before participating into this study. Our CT scan protocol was set up by the experienced radiologist to ensure the lowest effective radiation dose with the best image quality (less than 2 mSv). The acquisition had 384 joint slices with a 3 mm thickness and a matrix of 512 × 512 pixels (Figure 4.1(a)). The time duration of the full image acquisition was about 30 s.

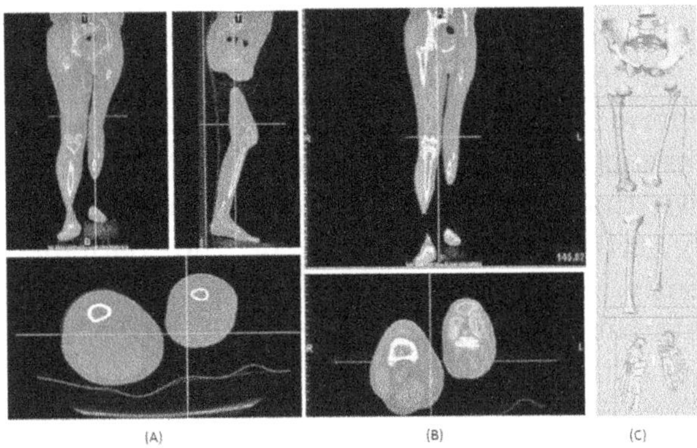

Figure 4.1. *a) CT-based anatomical images, b) threshold-based segmented images and c) STL-based skeletal surface envelop model of the lower limb structures of a PPRP patient*

Gait kinematics data have been collected on the PPRP patient with a KAFO attached on his left lower limb to support and compensate his skeletal deformity and muscle weakness. The KAFO was designed using a casting and subjective rectification process. The PPRP patient has no specific pain during his gait with KAFO. Note also that he cannot perform his gait without the KAFO. Another healthy subject (male, 39 years old, 175 cm height and 75 kg body mass) also participated in gait data acquisition to serve as reference kinetic data. Data acquisition was performed using the Helen Hayes protocol based on the Davis marker configuration model [DAV 91]. The 3D marker's trajectories were collected using a VICON Motion Capture system (Vicon, Oxford, UK) with six cameras. The 3D trajectories of the markers were filtered by a fourth-order Butterworth filter with cut-off frequency of 5 Hz (Vicon Nexus software,

Oxford, UK). Gait kinetics data such as foot–ground reaction forces were acquired using two force platforms AMTI (AMTI, Watertown, United Kingdom).

4.1.2.2. *Patient-specific biomechanical analysis of the effect of the KAFO*

On the basis of acquired 3D trajectories of the skin-based markers, joint kinematics during gait cycles including stance and swing phases of the PPRP patient were computed using a patient-specific biomechanical model developed with BGR.LifeMod software (LifeModeler, Inc., San Clemente, CA) [DAO 12]. This model consists of seven rigid segments (pelvis, thighs, legs and feet) (Figure 4.1(c)) derived from segmented CT images (Figure 4.1(b)) with 20 degrees of freedom [DAO 12]. 3D Slicer software (Surgical Planning Laboratory – MIT 1998) was used to perform segmentation of all rigid segments using the semi-automatic method. On the basis of 2D CT-based slice-by-slice anatomical images, bony segments were segmented from surrounding tissues. A threshold-based segmentation method was applied. Then, each image was verified and cleaned to ensure that the segmented pixels belonged to the tissue of interest as well as to avoid unnecessary segmented pixels, respectively. 3D geometries of all these segments were reconstructed and stored in STL format. Then, these geometries were used to develop a patient-specific osteoarticular model. The inverse kinematics (IK) algorithm was used to compute the joint kinematics using 3D trajectories of skin-mounted markers. It is important to note that the calibration between the model derived from CT images and 3D motion data is performed using an optimization approach provided by the BGR.LifeMod software. Thus, this process aims to put the supine imaging data and standing kinematics data into the same reference system. During the calibration step, the Euclidian distance errors between the coordinates of markers posed virtually on the image-based model and those of the real skin-mounted markers are minimized.

4.1.2.3. *Definition of evaluation parameters*

To quantify the change of kinematics patterns of the PPRP patient due to the effect of the KAFO, his joint kinematics were compared to those of the reference range of values of the normal subject extracted from reference data provided by the OpenSIM software [DEL 07] and standard deviation from literature-based values [MCG 09]. Absolute deviation of joint kinematics (angles in degree) was computed using the following formula:

$$dev^{kin}\left(^{\circ}\right) = a_i^{normal} - a_i^{pprp} \text{ with } a \in \left\{hip, knee, ankle\right\},\ i \in \left\{0,..,100\right\} \qquad [4.1]$$

where a is the hip or knee or ankle joint and i is the gait cycle percentage. Thus, mean absolute deviations and their standard deviation were computed during different phases of gait ranging from heel strike to terminal swing phases.

Concerning the analysis of kinetic patterns, relative peak-to-peak GRF deviations were computed at the heel strike (*hs*) and toe-off (*to*) during stance phase by using the following mathematical formula:

$$dev^{kinetics}\left(\%\right) = 100 \times \left(\frac{grf_j^{pprp} - grf_j^{normal}}{grf_j^{normal}}\right) \text{ with } j \in \left\{hs, to\right\} \qquad [4.2]$$

where GRF is the ground reaction force.

Relative contact time based on the GRF was also computed by using the following mathematical formula:

$$dev^{time}\left(\%\right) = 100 \times \left(\frac{\left(t_{to}^{s} - t_{hs}^{s}\right) - \left(t_{to}^{normal} - t_{hs}^{normal}\right)}{\left(t_{to}^{normal} - t_{hs}^{normal}\right)}\right) \text{ with } s \in \left\{affected, unaffected\right\} \qquad [4.3]$$

Data postprocessing was performed using Matlab R2008b (The Matworks, Inc., Natick, MA).

4.1.3. *Results*

4.1.3.1. *Kinematics analysis*

Hip joint kinematics of the PPRP patient were compared to those of the normal reference values (Figure 4.2). For the unaffected side of PPRP patient, we found that his hip kinematics pattern is quasi-similar to that of the normal range of values. Mean absolute deviation ranges from – 1 ± 0.2° to 9.3 ± 1° with maximal deviation at the toe-off phase (Tables 4.1 and 4.2). However, there is a phase shift between these kinematic patterns. Furthermore, the PPRP patient reaches the maximal extension amplitude peak faster than the normal subject. Concerning the affected side with orthosis of the PPRP patient, his joint kinematic pattern is completely altered by the orthosis effect. During the heel strike phase, the PPRP hip joint begins its flexion-to-extension activity with a significantly smaller angle than that of a normal range of values. Moreover, there is a redundant flexion activity during the swing phase (i.e. from 60% to 100% of gait cycle) of gait. We noted that the mean absolute deviation ranges from 0.1 ± 9.5° to –27.9 ± 3° with the maximal deviation at the heel strike phase (Table 4.1).

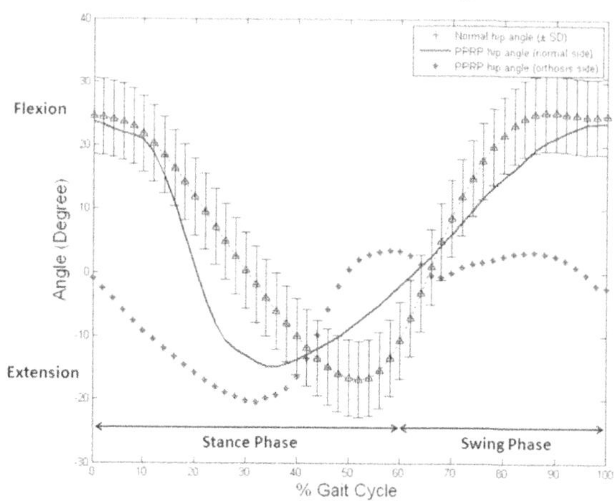

Figure 4.2. *Comparison of hip kinematics between PPRP data and normal reference data*

Joint		Joint kinematics deviation stance phase (0–60%)		
		Heel strike (0–2%)	Midstance (10–30%)	Toe-off (50–60%)
Hip(in ° ± SD)	Left side (with orthosis)	–26.3 ± 1	–27.9 ± 3	17.4 ± 2
	Right side	–1 ± 0.2	–8.2 ± 6	9.3 ± 1
Knee(in ° ± SD)	Left side (with orthosis)	-	-	-
	Right side	8.4 ± 3.2	–2 ± 3.5	40.9 ± 15.5
Ankle (in ° ± SD)	Left side (with orthosis)	10.5 ± 1.1	11.7 ± 1.3	7.5 ± 6.5
	Right side	3.5 ± 1.6	–2.3 ± 4.4	8.3 ± 3

Table 4.1. *Absolute mean deviation of joint kinematics of the PPRP patient comparing to the normal ranges of values during stance phase*

Joint		Joint kinematics deviation swing phase (60–100%)		
		Initial swing (60–73%)	Midswing (73–87%)	Terminal swing (87–100%)
Hip (in ° ± SD)	Left side (with orthosis)	0.1 ± 9.5	–17.6 ± 3.8	–24 ± 2
	Right side	2 ± 4.7	–5.8 ± 0.9	–2.8 ± 1.8
Knee (in ° ± SD)	Left side (with orthosis)	-	-	-
	Right side	–22.7 ± 21	–31.5 ± 8.9	2.2 ± 9.8
Ankle (in ° ± SD)	Left side (with orthosis)	15.1 ± 4.8	4 ± 1.5	5.2 ± 1.5
	Right side	10 ± 3	1.8 ± 2.1	1.6 ± 0.7

Table 4.2. *Absolute mean deviation of joint kinematics of the PPRP patient comparing to the normal ranges of values during swing phase*

Quasi-similar kinematics pattern between PPRP knee kinematics and the normal range of values was found for the unaffected side of the PPRP patient (Figure 4.3). However, the initial knee flexion is almost omitted. A phase shift is also noted, especially for the knee flexion pattern. Moreover, the PPRP patient performs knee flexion faster than the normal subject. Mean absolute deviation ranges from – 2 ± 3.5° to 40.9 ± 15.5° with the maximal deviation at the toe-off phase (Tables 4.1 and 4.2). Concerning the affected side with orthosis of the PPRP patient, knee angles were quasi-null during all gait cycles due to the locked joint of the KAFO at the knee level.

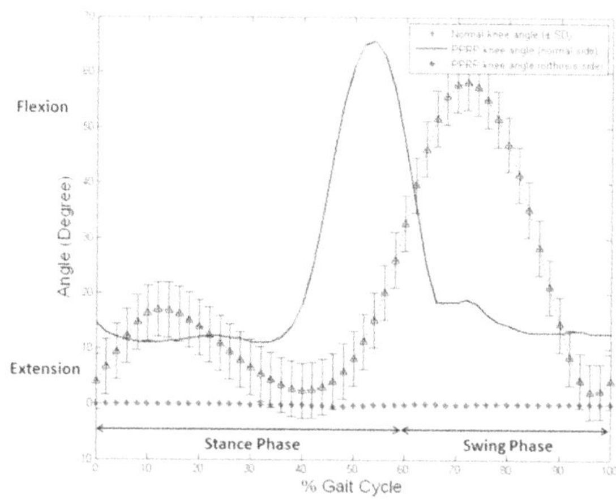

Figure 4.3. *Comparison of knee kinematics between PPRP data and normal reference data*

In the case of the ankle joint, the PPRP ankle kinematic pattern of the unaffected side is quasi-similar to that of the normal range of values. However, the ankle kinematic pattern is not smooth at the dorsiflexion and plantar flexion activites (Figure 4.4). The maximal dorsiflexion peak of the PPRP patient is greater than that of the normal range of values. The maximal plantar flexion peak of the PPRP patient is smaller than that of the normal range of values. Mean absolute deviation ranges from 1.8 ± 2.1° to 8.3 ± 3° with the maximal deviation at the toe-off phase (Table 4.1). In the affected side with orthosis, the ankle kinematics omitted the initial plantar flexion phase. The maximal dorsiflexion peak is reached faster than that of the normal range of values. Moreover, the ankle kinematics is not smooth during all gait cycles. Mean absolute deviation ranges from 4 ± 1.5° to 15.1 ± 4.8° with the maximal deviation at the initial swing phase (Table 4.1).

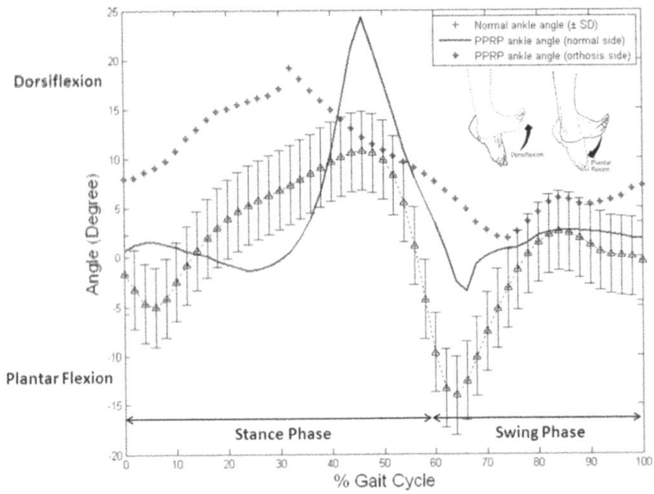

Figure 4.4. *Comparison of ankle kinematics between PPRP data and normal reference data*

4.1.3.2. *Kinetics analysis*

Experimental vertical GRF of PPRP patient compared to those of normal range of value are illustrated in Figure 4.5. We noted that there is a phase shift between PPRP kinetics patterns and those of the normal subject. Moreover, kinetic patterns of the PPRP patient are not smooth compared to those of the normal range of values. The PPRP patient reaches the heel strike and toe-off phases slower than the normal subject. Furthermore, the foot–ground contact time of the PPRP patient is longer than that of the normal subject, especially for his unaffected side. The relative peak-to-peak GRF revealed that the orthosis contributes to the increasing foot–ground reaction forces of 14% and 5% at the heel strike phase for the affected and unaffected sides, respectively. At the toe-off phase, PPRP's foot–ground reaction force increases (approximately 4.5%) for the affected side and it decreases (around –14.3 %) for the unaffected side. Moreover, we noted that the orthosis effect contributes to the increase in relative contact times of around 212% and 117% for the PPRP's affected and unaffected sides, respectively.

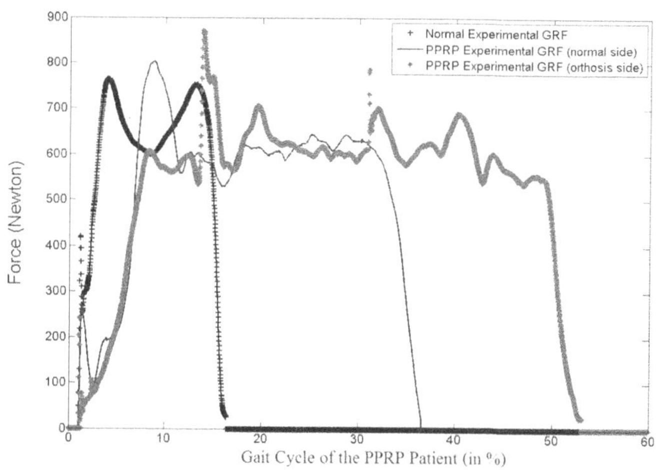

Figure 4.5. *Comparison of experimental ground reaction forces in the vertical direction between PPRP data and normal reference data during stance phase (60% of a gait cycle)*

4.1.4. *Discussion*

Musculoskeletal geometries and functions of the PPRP patient show abnormal characters consisting of skeletal deformities and muscle paralysis. So the qualitative and quantitative assessment of the consequence effect of the KAFO under the pathophysiological state of the PPRP patient on the biomechanical functions of this musculoskeletal disorder needs to be investigated in a patient-specific manner. In this study, we showed that the combination of medical imaging and 3D motion capture techniques provides a potential technical routine practice for the patient-specific biomechanical analysis of the effect of the KAFO on the PPRP patient during motions such as gait. Moreover, a PPRP patient-specific KAFO design could be established to optimize its biomechanical recovering functions.

In fact, joint kinematics patterns at the hip, knee and ankle joints of the PPRP patient during his gait were altered due to the knee joint fixed by KAFO design as illustrated in Figures 4.2–4.4. Consequently, the gait kinematics on the PPRP's unaffected side is influenced. From a biomechanical point of view, this imbalance effect may have consequences on the mechanical properties of bones, muscles and osteoarticular joints. To avoid this imbalance loading effect on the joints and gait, it would be of importance to optimize the KAFO.

For the PPRP's hip kinematics, the use of orthosis creates a phase shift for both sides as well as an additional and longer flexion activity for the affected side during its swing phase of gait (from 60% to 100%). The improvement of the hip joint pattern could be performed during the transition extension-to-flexion phase to avoid the additional flexion activity for the affected side. In this case, the following objective function could be used to optimize the orthosis performance at the hip joint:

$$Minimize\ f_{hip}^{kinematics} = \sum \left| angle_i^{normal\_hip} - angle_i^{pprp\_hip} \right|, \quad \forall i \in \{0,..,100\} \qquad [4.4]$$

For the PPRP's knee kinematics, the unaffected side has a phase shift and a faster flexion peak while the affected side is completely locked due to the KAFO's functional fixed joint. The results suggest that the KAFO design should allow joint movement such as the flexion/extension activities of the knee during gait. This should help decelerate the flexion velocity of the PPRP's unaffected side (Figure 4.3). For this joint, the following objective function could be used to optimize the orthosis performance at the knee joint:

$$Minimize\ f_{knee}^{kinematics} = \Sigma \left| angle_j^{normal\_knee} - angle_j^{pprp\_knee} \right|,\ \forall j \in \{0,..,100\} \quad [4.5]$$

For the PPRP's ankle kinematics, there are phase shifts over times as well as over joint amplitudes. Moreover, the kinematic patterns are not smooth at all gait cycles, especially at the maximal dorsiflexion and plantar flexion and in the dorsiflexion-to-plantar flexion transition. Furthermore, foot–ground reaction forces reveal that the orthosis decelerated the foot–ground contact time for both affected and unaffected sides. The patterns of foot–ground reaction forces are altered leading to the lack of three separate contact peak forces at heel strike, stance and toe-off phases. Consequently, an ankle joint design allowing smoother ankle kinematic patterns should be considered. In this case, following objective functions could be used to optimize the orthosis performance at the ankle joint:

$$Minimize\ f_{ankle}^{kinematics} = \Sigma \left| angle_k^{normal\_ankle} - angle_k^{pprp\_ankle} \right|,\ \forall k \in \{0,..,100\} \quad [4.6]$$

$$Minimize\ f_{foot}^{kinetics} = \Sigma \left| GRF_l^{normal\_foot} - GRF_l^{pprp\_foot} \right|,\ \forall l \in \{0,..,100\} \quad [4.7]$$

In summary, based on the kinematics and kinetics patterns of the PPRP patient from patient-specific biomechanical analysis, a design-oriented multiobjective function could be used to optimize the KAFO design of PPRP patients as follows:

$$Minimize\ f_{orthosis}^{design} = \Sigma \left(w_1 f_{hip}^{kinematics} + w_2 f_{knee}^{kinematics} + w_3 f_{ankle}^{kinematics} + w_4 f_{foot}^{kinetics} \right) \quad [4.8]$$

where w_1, w_2, w_3, w_4 are the kinematics and kinetics weighted design factors. These factors will be adapted to the specificity of the target orthopedic disorder.

The use of 3D motion capture in the quantitative assessment of the medical support devices such as orthosis or prosthesis is not new [JOH 04, KAR 11, SIL 11]. However, only generic and simplified biomechanical models (e.g. VICON or visual 3D) were used to perform this task leading to inaccurate analysis results [CAP 05, CRO 05, FER 08]. Our study opens a new direction in which to perform patient-specific biomechanical analysis of the effect of medical orthopedic devices using the coupling between medical imaging and 3D motion capture, thereby leading to more accurate analysis results. In fact, the use of patient-specific models could lead to better evaluation and design characteristics than the use of generic models. The first advantage relates to the individualized bone geometries, which allows a more accurate joint range of motion to be acquired. The second advantage is the possible individualized design of the orthosis using individualized geometries derived from medical imaging and numerical approach such as virtual computer-aided orthosis design [MAV 11]. Furthermore, on the reliable analysis results, objective design specifications could be recommended to optimize the performance of the orthopedic devices [JOH 04]. Our analysis was focused only on the kinematics and kinetics data in the sagittal plane because of their reliability as shown in previous study [DAO 12].

From a methodological point of view, the main limitation of the present methodology remains in the fact that the development of *in silico* patient-specific models derived from medical imaging and 3D motion capture requires advanced knowledge and expertise in image processing and physics-based simulation. Another limitation is the time-consuming character of the development of an image-based model. However, with the progress of methodological and technical approaches, this development process has been standardized. In fact, different modeling and computing steps have been optimized leading to facilitate their developments as well as to reduce the processing time for a clinical purpose [CHA 03, BLE 07].

4.2. Computational musculoskeletal ontological model

4.2.1. *Introduction*

The connection of knowledge is necessary in all the scientific fields, but the diversity of representation of knowledge is an obstacle to formalizing research. Ontology appeared as a good solution because it gives the possibility of sharing the common comprehension of the structure of information between researchers [THO 07, STE 07]. Then, it also allows the reuse of the knowledge in various computer-based systems. This approach is also applied to building the knowledge-based system using the accumulated knowledge in a machine-readable format to reason, diagnose or give adequate decisions [DAN 06]. In medicine, the exponential increase of biomedical data and knowledge has also led to the application of ontological methods for reusing the voluminous knowledge.

The ontology is defined as an explicit specification of a conceptualization of a field. The conceptualization is also the abstract model of real phenomena. To facilitate the comprehension by keeping the principal idea, we propose our definition of the ontology as a structured specification of the concepts and semantic, structural, functional and intelligent relationships in a field [DAO 07]. The quality of ontologies depends on the degree of representation of a certain portion of reality [SCH 03].

Biomechanics has recently come into full evolution with many research topics performed in computer-based simulation of the musculoskeletal system [DAO 12, BLE 07, DEL 07]. An inevitable problem is the heterogeneity of research studies. Moreover, developed models are simplified and especially partial. There is a need to structure the vast amount of conceptual knowledge in this field in order to access to biomechanics data such as controlled vocabularies, terminologies and thesauri. In the literature, there are many studies concerning biomedical ontologies [ZWE 95, CHA 06, ROS 03, HUN 06]. The Open Biomedical Ontologies (OBO) project has also been developed to enrich biological fact databases [SMI 07, DIN 07]. A large amount of terminologies of the biological structure have been published. The Foundational Model of Human Anatomy has already been represented [ROS 03]. But there is no study on biomechanics ontology and the diagnosis of the pathologies of the lower limbs. We have recently initiated it for the musculoskeletal system. The aim of this section is to present a computational ontology of the musculoskeletal system of the lower limbs.

4.2.2. *Materials and methods*

4.2.2.1. *Representation formalism*

We use the platform Protégé 2000 to develop our ontology using the OWL formalism [RUB 07]. OWL is a standard ontology markup language based on the DLs for the Semantic Web. DLs are expressed by a formal, logic-based semantics and they are used for different ontologies. The Protégé 2000 also permitted us to facilitate creating and reasoning about ontologies through a graphical user interface.

4.2.2.2. *The semantic-based extraction process of the knowledge*

In the knowledge representation literature, many general methodologies for developing ontologies have been described [STA 01, ALE 06]. The knowledge of the ontology could be extracted manually or semi-automatically or automatically from the raw text. This approach is not efficient due to the requirement of a powerful extraction tool and a representative text base. To face this disadvantage, we propose a semantic-based extraction approach process and its application to construct a diagnostic-based ontology. The top-down strategy is used to identify concepts and relationships between concepts. We started at the general concepts and then identified the specific concepts. For example, the bone concept is verified before the tibia concept. This semantic-based extraction process of the knowledge (Figure 4.6) is based on the partitions and granularity theory [BIT 03]. In this approach, the objects and relations are made more easily graspable by cognitive subjects. The input is the expertise of the biomechanical experts and this process begins with the interview and the discussion with the experts. The biomechanical experts work directly with the knowledge engineer to treat, normalize and verify the semantics of the knowledge. This process is iterative to verify directly the semantics of the knowledge with the experts. In the semantic-based extraction, the graph construction and the document ontology construction are considered. This semantic-based extraction consists of the following steps: (1) enumerate the important terms; (2) define the classes and the hierarchy of these classes; (3) define the properties of the classes-attributes; (4) define the facets of the attributes: cardinality, types of value, domain, etc.; (5) create the instances. This semantic-based extraction process is specified, dedicated to the construction of the diagnostic-based ontology where the physiological and functional semantics decide the robustness of the system. The uncertainties are guaranteed by an iterative verification unit.

4.2.2.3. *Principal software components*

First, the conceptual structure of the OSMMI has to be established by using the Unified Modeling Language (UML) schema in order to have a general sight of ontology. Second, we start to create the OSMMI using the platform Protégé 2000 (Stanford University). On the basis of the built ontology, the last component is the construction of the reasoning part (Figure 4.7).

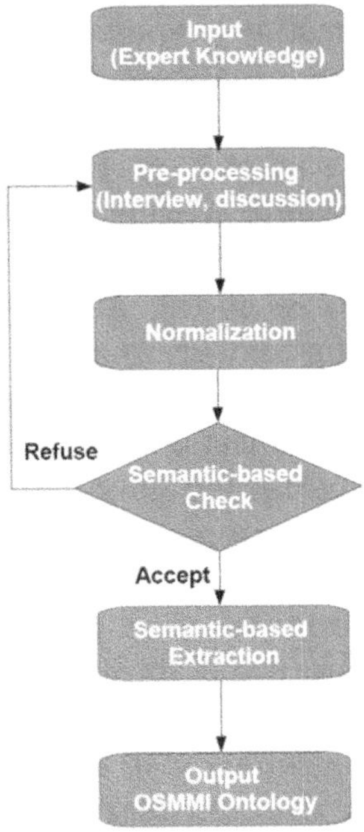

Figure 4.6. *The semantic-based extraction process of the knowledge*

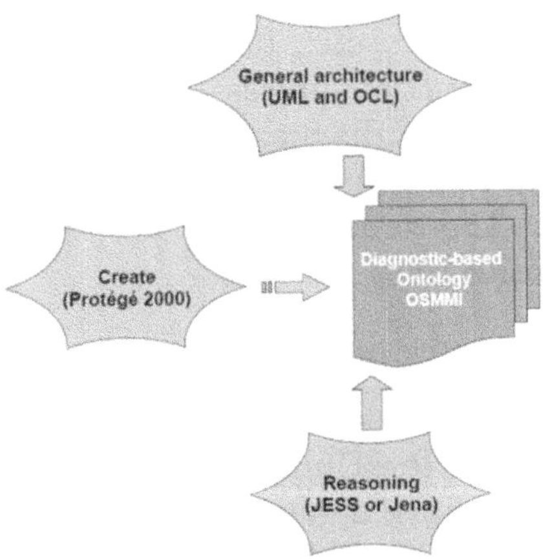

Figure 4.7. *Principal software components concerning the musculoskeletal ontology*

4.2.3. *Results*

4.2.3.1. *Ontology*

The architecture of our ontology includes 14 functional and anatomical structures of the musculoskeletal system of the lower limbs (Figure 4.8 and Table 4.3), which are defined like classes of ontology: nervous system, ligament, muscle, tendon, cartilage, bone, limb, posture, support of load, diarthrosis joint, movement, articular contact, contact of environment and gait.

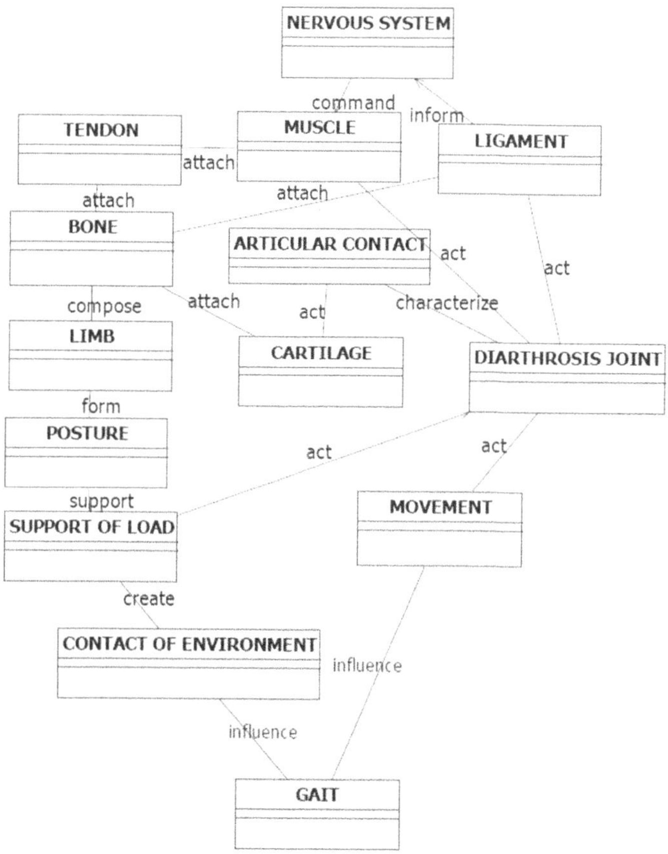

Figure 4.8. *Principal concepts of the musculoskeletal ontology*

Relation	Inverse relation	Description
Inform	N/A	Unidirectional relationship
Command	N/A	Unidirectional relationship
Attach	Inverse of attach	Bidirectional relationship
Compose	Inverse of compose	Bidirectional relationship
Act	Inverse of act	Bidirectional relationship
Influence	Inverse of influence	Bidirectional relationship
Form	Inverse of form	Bidirectional relationship
Support	Inverse of support	Bidirectional relationship
Create	Inverse of create	Bidirectional relationship
Characterize	Inverse of characterize	Bidirectional relationship

Table 4.3. *Musculoskeletal ontology: class-level relations between anatomical and functional constituents*

For the physiological and functional semantics of our ontology, what is considered as implicit on anatomy and function is defined by the following relationships:

– inform: the ligaments inform the nervous system if there are active signals;

– command: the nervous system commands the muscles;

– attach: the muscles are attached to the bones through the tendons; the cartilage and the ligaments are attached to the bones;

– compose: the limb is composed of different bones that correspond to a particular function in gait;

– act: the muscles, the support of load, the movement and the ligaments act on the diarthrosis joint; the cartilage acts as the articular contact;

– influence: the contact of environment, the movement influence the gait;

– form: the limbs form the corresponding posture;

– support: the posture is supported by the support of load;

– create: the support of load is created by the contact with the environment;

– characterize: the articular contact characterizes the diarthrosis joint.

All parameters are collected properly for each part of the musculoskeletal ontology. These parameters are well chosen to study the impact of the pathologies of the lower limbs. They are considered as follow:

– Nervous system: to decrease the complexity of our ontology, we do not take into account the characteristics of the nerve control, and we define a variable state having two states: active or inactive.

– Bone: we define the length, size, weight, center of mass, proximal diameter, distal diameter, orientation alpha, orientation beta, orientation delta, elasticity, shearing, density and ultimate stress.

– Tendon: we define the length, size, weight, center of mass, proximal diameter, distal diameter, orientation alpha, orientation beta and orientation delta.

– Muscle: we define the length, size, weight, center of mass, proximal diameter, distal diameter, orientation alpha, orientation beta, orientation delta, elasticity, force, strength, physiological cross-sectional area, maximum tissue stress, resting load and overall muscle tone.

– Ligament: we define the length, size, weight, center of mass, proximal diameter, distal diameter, orientation alpha, orientation beta, orientation delta and elasticity.

– Cartilage: we define the length, size, weight, center of mass, proximal diameter, distal diameter, orientation alpha, orientation beta, orientation delta and elasticity.

– Movement: we define the position, the velocity and the acceleration.

– Contact with the environment: we define the generated force by the environment, the reaction force at ground contact.

– Support of load: we define the load.

– Limb: we define the number and the name of bones that compose the limb of interest.

– Posture: we define the limbs that form the posture of interest, and also the state of current posture: normal or abnormal.

– Articular contact: we define the pressure, the zone of contact.

– Diarthrosis Joint: we define the force, the moment and the torque.

– Gait: we define the average velocity, the hip, knee and ankle torques, the GRF, the level of balance, and the intersegmental angles and forces.

We define the semantic relationships of our ontology such as the object properties and the parameters are defined such as the data-type properties in the Protégé framework (Table 4.4).

Classes	DP	OP	Restriction
119	100	31	131

Table 4.4. *Musculoskeletal ontology: number of defined classes, data properties (DP), object properties (OP) and restrictions*

For each part of our ontology, there are several subparts (as illustrated in Figure 4.9). We define the elements needed for the gait analysis. The hierarchy of our ontology is defined as follows:

– Bone: we define the subclasses of bone such as the talus, the calcaneus, the first cuneiform, the second cuneiform, the third cuneiform, femur, tibia, fibula, iliac, ilion, ischion, kneecap, metatarsal bone, distal phalanges, intermediate phalanges, proximal phalanges, pubis and scaphoide.

– Muscle: we define 22 principal muscles: the adductor magnus, biceps crural, crural, gluteus maximus, gluteus medius, gluteus minimus, lateral peroneus brevis, lateral peroneus longus, long biceps, calf, phyramidal, quadriceps, rectus femoris, sartorius, semimembranosus, semitendinosus, short biceps, soleus, tensor faciale latae, tibialis anterior, tibialis posterior and the triceps sural.

– Ligament: we chose the ligaments needed for gait analysis such as the iliofemoral, posterior ilio femoral, ischio femoral, lateral ligaments, anterior ligament, posterior ligament (which has two subclasses: arched poplity and oblique poplity), patellar aileron (which has two subclasses: patellar external aileron and patellar internal aileron), pubofemoral, anterior cross-ligament, posterior cross-ligament, internal ligament, external ligament, anterior tibio-fibular, posterior tibio-fibular, ichio-femoral and pubo-femoral.

– Tendon: we define the following subclasses of the tendons: the gastroneminus and the plantaris tendons.

– Articular contact: we define the following subclasses: the contact between the cartilage and the cartilage, and the contact between the cartilage and the meniscus.

– Diarthrosis joint: we define the following subclasses: the coxo-femoral joint, femoro-tibial, patello-femoral, sacro-iliaque, symphyse pubic and tibiotarsienne.

– Cartilage: we define the following subclasses: the hyaline cartilage, elastic cartilage and fibrocartilage.

– Movement: we define the following subclasses: the translation and the therotation.

– Limb: we define the following subclasses: the pelvis, thigh, leg and foot.

– Contact with the environment: we define the following subclass: the foot–floor contact.

– Posture: we define the following subclasses: the neutral, the varus and the valgus.

– Gait: we define the following subclasses: the normal gait, the crouch gait, rotational abnormality (RA) and cerebral palsy.

– Support of load: we define the following subclasses: the hip load, the knee load and the ankle load.

Figure 4.9. *OSMMI ontology (in OWL format). Classes of the physiological and functional structures are shown on the left panel, and formal definitions of the structure are shown on the right. The class Tibia is defined by five assertions, all necessary conditions for this class. The disjoint classes of the Tibia are also defined*

4.2.3.2. *Application*

As an example, the RA [CAH 92] is studied as the pathological case to test reasoning aspect of the musculoskeletal ontology. It is important to note that this pathology is one of the most frequent consultations in pediatric orthopedics [CAH 92]. Different causes can be deduced by the biomechanics experts:

– position (directions: lateral, medial, frontal, symmetrical, asymmetrical) of the knee during stance phase;

– position (directions: lateral, medial, frontal, symmetrical, asymmetrical) of the foot during stance phase;

– relationship between femoral anteversion (FA) and medial hip rotation.

On the basis of these causes expressed above and by using the semantics of our musculoskeletal ontology, the RA could be derived from joint problem. Then, the schema would be in accordance with Figure 4.10 [Gait → Movement → Diarthrosis joint → Articular contact → Cartilage → Bone]. Another schema could be in accordance with Figure 4.11 [Bone → Limb → Posture → Support of load → Contact of environment → Gait]. These results will be used in orthopedics treatment; when the physiotherapist realizes a functional rehabilitation gesture, these above relationships will be taken into account to avoid impairment of the penetrating injuries of the anatomical and functional parts of the musculoskeletal system of the lower limbs. For example, before the treatment on the diarthrosis joint, the articular contact must be taken into account (first schema). Detailed anatomical and functional knowledge are only available to an expert but they can be encoded in ontologies and exploited by the computer applications to reason, diagnose or give adequate decisions. By applying the reasoning methods, new knowledge could be inferred from available knowledge asserted in the ontology.

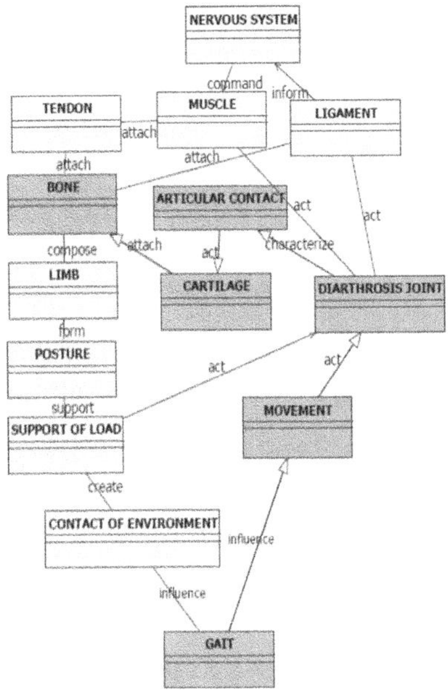

Figure 4.10. *Backward chaining scenario*

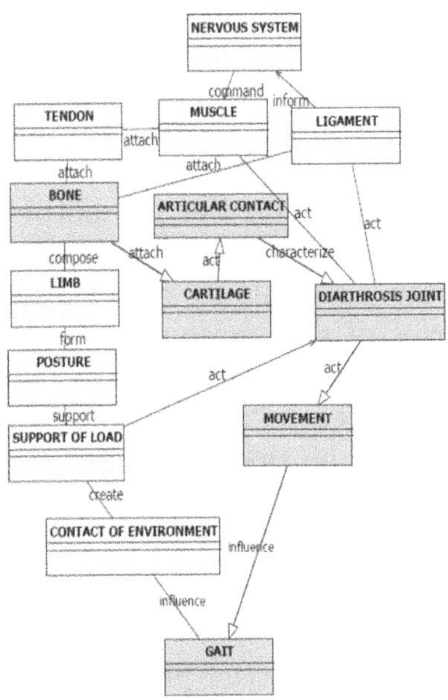

Figure 4.11. *Forward chaining scenario*

4.2.4. *Discussion*

The current musculoskeletal ontology consists of physiological and functional structures concerning the musculoskeletal system of the lower limbs. But the other structures such as the upper limbs or the nervous system can be easily developed and integrated into the ontology by using our semantic-based extraction process. The UML schema is extensible and the Protégé 2000 framework technically ensures the extensibility and the scalability of the ontology. The Protégé ontology editor permits the handling of semi-automated detection and handling of changes in ontologies. It is a user–friendly editor and our ontology is more accessible for non-specialist users. With the format OWL, our ontology is consulted even if someone does not have Protégé.

The developed ontology is a domain ontology and it is already permitted to link to medical top-level ontologies. Top-level ontologies describe the general concepts or categories that are presumed to be common across domains [ALE 06]. Many biomedical ontologies (domain or top-level ontologies) have been published: the Foundational Model of Anatomy [ROS 03], the GALEN Common Reference Model to provide the reuse of terminology resources for clinical systems [REC 91], the Gene ontology for representation of gene product information in different databases [GEN 01], the Medical Entities Dictionary [CIM 00], the SNOMED CT system, the DOLCE, the Cyc's upper ontology, the Sowa's top-level ontology and the UMLS Semantic Network.

The developed ontology has been used in many real applications such as the development of a knowledge-based search engine for the web-based Human Musculoskeletal System Resources (HMSR) [DAO 13] or development of an ontology-driven decision support system for gait analysis

[TUR 13] or the development of an ontology-based modeling pipeline to create specific musculoskeletal simulation [DIC 13].

Protégé is a powerful ontology development environment (http://protege.stanford.edu). This platform is dedicated to biomedical applications. It is supported by a great community of developers and users. It is easy to share our common understanding of information among people or intelligent agents. It provides functionalities for editing classes, properties and instances. The knowledge is accessed by locating the pertinent anatomical and functional entities (classes in our ontology), and recovering their attributes (slots on the class). By using this platform, the developed musculoskeletal ontology could be shared and reused easily in the fields of biomechanics and biomedical informatics.

4.3. Predictive models of the pathologies of the lower limbs

4.3.1. *Introduction*

The period of 2000–2010 was labeled as the "Decade of Bones and Joints". The pathologies of the musculoskeletal system of lower limbs are various and complex. These pathologies directly influenced the quality of life of the people involved [CAH 92, ARN 05]. The diagnosis of these pathologies aims to depict the abnormal observations describing the altered behavior of the anatomy or function of the human body. The diagnosis is based on the clinical data of the patient. These data are heterogeneous and multidimensional. Thus, the interpretation of these data is not obvious. It could be of great clinical interest if an automatic process could be proposed to extract significant pathological parameters in term of probability.

Classification is an important task in a data mining application that exploits large amounts of data and picks out relevant information. The increasing number of electronic medical database has led to the study of data mining techniques for building classification models [HAN 06]. The classification method estimates the risk factors by computing the most significant parameters in terms of probability based on the information gain and entropy (decision tree), the support vectors (support vector machines) and the error propagation (artificial neural network).

The objective of this section is to study some classical data mining techniques to develop predictive models of the pathologies of the lower limbs. Then, a specific case of clubfoot deformities [DAO 11] will be analyzed.

4.3.2. *Materials and methods*

4.3.2.1. *Data sets*

The first studied pathology of the lower limbs is the RA (Figure 4.11). A total of 15 morphological properties (Q angle, lateral and medial rotation angles at the hip joint, tibial torsion angle, femoral anteversion angle, etc.) of the RA are included into the predictive models. The symbolic data are used as training and testing data sets. The training data set includes 100 cases and the testing data set includes 70 cases. The cases are randomly selected.

The second studied pathology is the clubfoot deformities (Figure 4.12). A total of 15 morphological parameters (e.g. midfoot supination angle, equinus angle and varus angle) are used to perform the classification model. The training data set includes 700 random cases and the testing data set includes 300 random cases.

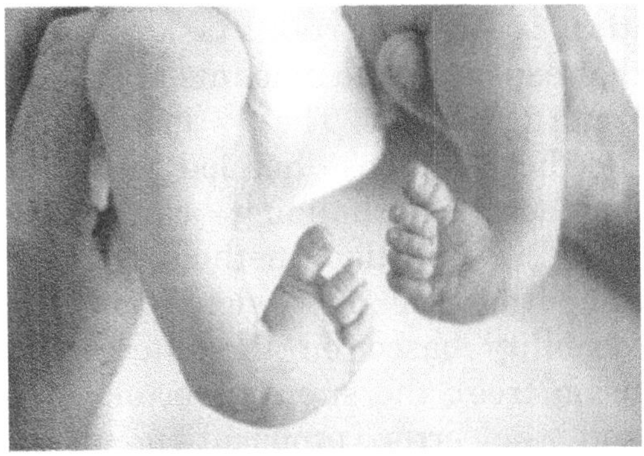

Figure 4.12. *Clubfoot deformities (Robert Debré, Paris, France)*

4.3.2.2. *Data mining methods*

The classical decision tree, support vector machine and artificial neural network are used to develop predictive models of the rotational abnormalities. The C4.5 decision tree and random forest methods are used. Two types of neural network algorithms are applied as RBF neural network and MLP neural network. Different kernel functions (polynomial, Gaussian) of the support vector machine method are tested.

The decision tree was used to develop the classification model of the clubfoot deformities. Then, a user interface was created allowing the clinician to make the automatic classification using the patient data.

4.3.3. *Results*

4.3.3.1. *Explicit knowledge model of the rotational abnormalities*

The knowledge model induced by the classical decision tree method is shown in Figure 4.13. It leads to the observation that the Q angle and tibial torsion angles are the most significant and discriminant parameters for depicting the rotational abnormalities. The RA pathology rule set could be deduced from the decision tree and expressed as follows: (1) If Q – angle > 17, then class = NRA; (2) If Q – angle ≤ 17 and ttc> 10, then class = RA; and (3) If Q – angle ≤ 17 and ttc ≤ 10, then class = NRA.

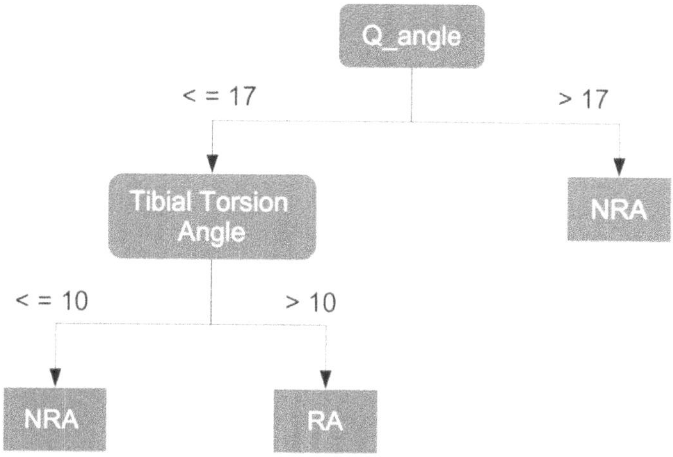

Figure 4.13. *Decision tree model of the rotational abnormalities (RA): NRA means the normal case*

4.3.3.2. *User interface of the classification system of the clubfoot deformities*

The user interface of the clubfoot classification model was developed as a web-based interface (Figure 4.14). The diagnosis, treatment and monitoring of the clubfoot deformities were set up via a step-by-step process. The validation of clubfoot deformities' application was performed using real patient data from hospital of Robert Debré, Paris, France.

4.3.3.3. *Performance comparison of different data mining methods*

The RBF neural network gives better results than the MLP neural network (1% versus 5.9% error rate) (Figure 4.15). The error rate is defined as the percent of the poorly-classified pathological cases. Among the SVM methods, the polynomial kernel methods (1% error rate) give better results than the Gaussian kernel methods. The polynomial kernel degree 2 & 3 converged to the optimal classifications with 1% error rate. But among these methods, the decision tree gives the best results for the error rate (1%) and also for the computing time (<50 ms).

Figure 4.14. *User interface of the classification system for the clubfoot deformities*

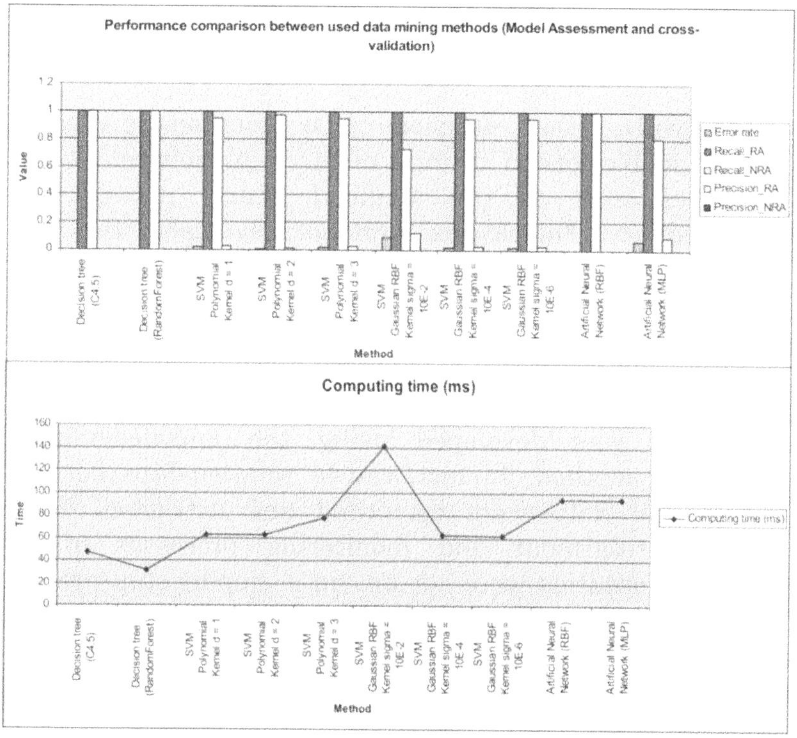

Figure 4.15. *Performance comparison of different data mining methods for RA instances*

4.3.4. *Discussion*

In the case of an RA study, we found that the decision tree with C4.5 algorithm gives a good performance for the trained and tested instances. Furthermore, the results of the C4.5 are more comprehensive in terms of result explanation for the medical diagnostic. As a result, the rule set is reported to determine risk factors and the effect of these factors on the pathological case. In the literature, the comparison between decision trees and support vector machines is studied [LOR 07]. In this study, the comparison of three methods (decision tree, artificial neural network and SVM) was studied. The results showed that these methods converge to optimal classifiers for the RA pathology with a 1% error rate.

Specifically, the decision tree method gives an understandable format in form of the reduction rule. Other approaches such as the artificial neural network are considered as a "black box" and not obvious to the user. Consequently, the decision tree could be a good choice for the medical decision-making in the framework of *in silico* medicine.

For the clubfoot deformities, a classification model as a meta-model was developed using the knowledge-based engineering method. In fact, causal relationships between experimental data were computed and extracted. Then, diagnosis, treatment and monitoring processes of the clubfoot deformities were set up via a step-by-step process. Patient information and individual development states of the patients were centralized. This reduces the medical cost as well as the administration time [DAO 11].

4.4. Conclusions

In this chapter, some applications of biomechanical and knowledge-based models are expressed and analyzed. However, to reach the clinical objectives, these models need to be analyzed further with larger populations. Moreover, the promotion of these models in the clinical routine practice needs a strict collaboration between biomechanical experts and clinicians. Furthermore, methodological and technical aspects need to be investigated to improve the precision and the accuracy of current biomechanical and knowledge-based models.

4.5. Summary

– The biomechanical musculoskeletal model could be used to study the effect of the orthosis on the mechanical function of the human body.

– The musculoskeletal ontology allows the knowledge of the musculoskeletal system to be formalized and reused by human being or computers.

– Data mining methods could be used to develop predictive models of the pathologies of the lower limbs.

– The decision tree provides an easy-to-understand predictive model for the clinical interpretation of medical observations.

4.6. Bibliography

[ALE 06] ALEXANDRE C.Y., "Methods in biomedical ontology", *Journal of Biomedical Informatics*, vol. 39, no. 3, pp. 252–266, 2006.

[ARN 05] ARNOLD A.S., DELP S.L., "Computer modeling of gait abnormalities in cerebral palsy: application to treatment planning", *Theoretical Issues in Ergonomics Science*, vol. 6, nos. 3–4, pp. 305–312, 2005.

[BIT 03] BITTNER T., SMITH B., "A theory of granular partitions", *Foundation of Geographic Information Science*, pp. 117–151, 2003.

[BLE 07] BLEMKER S.S., ASAKAWA D.S., GOLD G.E., *et al.*, "Image-based musculoskeletal modeling: applications, advances, and future opportunities", *Journal of Magnetic Resonance Imaging*, vol. 25, no. 2, pp. 441–451, 2007.

[BLE 08] BLEYENHEUFT C., CATY G., LEJEUNE T., *et al.*, "Assessment of the Chignon® dynamic ankle-foot orthosis using instrumented gait analysis in hemiparetic adults", *Annales de Readaptation et de Medecine Physique*, vol. 51, no. 3, pp. 154–160, 2008.

[CAH 92] CAHUZAC J.P., HO BA THO M.C., BAUNIN C., *et al.*, "Classification of 125 children with rotational abnormalities", *Journal of Pediatric Orthopaedics Part B*, vol. 1, pp. 59–66, 1992.

[CAP 05] CAPPOZZO A., CROCE U.D., LEARDINI A., *et al.*, "Human movement analysis using stereophotogrammetry: part 1: theoretical background", *Gait & Posture*, vol. 21, no. 2, pp. 186–196, 2005.

[CHA 03] CHAO E.Y.S., "Graphic-based musculoskeletal model for biomechanical analyses and animation", *Medical Engineering & Physics*, vol. 25, no. 3, pp. 201–212, 2003.

[CHA 06] CHARLET J., BACHIMONT B., JAULENT M.C., "Building medical ontologies by terminology extraction from texts: an experiment for the intensive care units", *Computer in Biology and Medicine*, vol. 36, nos. 7–8, pp. 857–870, 2006.

[CHE 05] CHEN G., PATTEN C., KOTHARI D.H., *et al.*, "Gait differences between individuals with post-stroke hemiparesis and non-disabled controls at matched speeds", *Gait & Posture*, vol. 22, no. 1, pp. 51–56, 2005.

[CIM 00] CIMINO J.J., "From data to knowledge through concept-oriented terminologies: experience with the Medical Entities Dictionary", *Journal of the American Medical Informatics Association*, vol. 7, no. 3, pp. 288-297, 2000.

[CRO 05] CROCE U.D., LEARDINI A., CHIARI L., *et al.*, "Human movement analysis using stereophotogrammetry: part 4: assessment of anatomical landmark misplacement and its effects on joint kinematics", *Gait & Posture*, vol. 21, no. 2, pp. 226–237, 2005.

[CRU 09] CRUZ T.H., DHAHER Y.Y., "Impact of ankle-foot-orthosis on frontal plane behaviors post-stroke", *Gait & Posture*, vol. 30, no. 3, pp. 312–316, 2009.

[CUL 09] CULLELL A., MORENO J.C., ROCON E., *et al.*, "Biologically based design of an actuator system for a knee–ankle–foot orthosis", *Mechanism and Machine Theory*, vol. 44, no. 4, pp. 860–872, 2009.

[DAN 06] DANIEL L.R., OLIVIER D., YASSER B., *et al.*, "Usingontologies linked with geometric models to reason about penetrating injuries", *Artificial Intelligence in Medicine*, vol. 37, pp. 167–176, 2006.

[DAO 07] DAO T.T., MARIN F., HO BA THO M.C., "Ontology of the musculoskeletalsystem of the lower limbs", *Proceedings of the 29th Annual International Conference of the IEEE Engineering in Medicine and Biology Society*, pp. 386–389, 2007.

[DAO 11] DAO T.T, MARIN F., BENSAHEL H., *et al.*, "Computer-aided decision system applied to the clubfeet deformities", *Advances in Experimental Medicine and Biology*, vol. 696, no. 7, pp. 623–635, 2011.

[DAO 12] DAO T.T., MARIN F., POULETAUT P., *et al.*, "Estimation of accuracy of patient specific musculoskeletal modeling: case study on a post-polio residual paralysis subject", *Computer Method in Biomechanics and Biomedical Engineering*, vol. 15, no. 7, pp. 745–751, 2012.

[DAO 13] DAO T.T., HOANG T.N., TA X.H., *et al.*, "Knowledge-based personalized search engine for the web-based human musculoskeletal system resources (HMSR) in biomechanics", *Journal of Biomedical Informatics*, vol. 46, no. 1, pp. 160–173, 2013.

[DAV 91] DAVIS R.B., OUNPUU S., TYBURSKI D., *et al.*, "A gait analysis data collection and reduction technique", *Human Movement Science*, vol. 10, pp. 575–587, 1991.

[DEA 95] DEAN E., AGBOATWALLA M., DALLIMORE M., *et al.*, "Poliomyelitis: Part 2: revised principles of management", *Physiotherapy*, vol. 81, no. 1, pp. 22–28, 1995.

[DEL 07] DELP S.L., ANDERSON F.C., ARNOLD A.S., *et al.*, "OpenSim: open-source software to create and analyze dynamic simulations of movement", *IEEE Transactions on Biomedical Engineering*, vol. 54, no. 11, pp. 1940–1950, 2007.

[DES 06] DESLOOVERE K., MOLENAERS G., GESTEL L.V., *et al.*, "How can push-off be preserved during use of an ankle foot orthosis in children with hemiplegia? A prospective controlled study", *Gait & Posture*, vol. 24, no. 2, pp. 142–151, 2006.

[DIC 13] DICKO A., GILLES B., FAURE F., *et al.*, "From generic to specific musculoskeletal simulations using an ontology-based modeling pipeline", *Studies in Computational Intelligence*, vol. 441, pp. 227–242, 2013.

[DIN 07] DINAKARPANDIAN D., TONG T., LEE Y., "A pragmatic approach to mapping the open biomedical ontologies", *International Journal of Bioinformatics Research*, vol. 3, no. 3, pp. 341–365, 2007.

[EUN 12] EUNSON P., "Aetiology and epidemiology of cerebral palsy", *Paediatrics and Child Health*, vol. 22, no. 9, pp. 361–366, 2012.

[FAR 10] FARAJ A.A., "Poliomyelitis: orthopaedic management", *Current Orthopaedics*, vol. 20, no.1, pp. 41–46, 2006.

[FER 08] FERRARI A., BENEDETTI M.G., PAVAN E., *et al.*, "Quantitative comparison of five current protocols in gait analysis", *Gait & Posture*, vol. 28, no. 2, pp. 207—216, 2008.

[GAR 06] GARRALDA M.E., MUNTONI F., CUNNIFF A., *et al.*, "Knee–ankle–foot orthosis in children with duchenne muscular dystrophy: user views and adjustment", *European Journal of Paediatric Neurology*, vol. 10, no. 4, pp. 186–191, 2006.

[GEN 01] GENE ONTOLOGY CONSORTIUM, "Creating the gene ontology resource: design and implementation", *Genome Research*, vol. 11, no. 8, pp. 1425–1433, 2001.

[GEN 10] GENET F., SCHNITZLER A., MATHIEU S., *et al.*, "Orthotic devices and gait in polio patients", *Annals of Physical and Rehabilitation Medicine*, vol. 53, no. 1, pp. 51–59, 2010.

[HAN 06] HAN J., KAMBER M., *Data Mining: Concepts and Techniques*, The Morgan Kaufmann Edition, 2006.

[HUN 06] HUNTER L., COHEN K.B., "Biomedical language processing: what's beyond PUBMED?", *Molecular Cell,* vol. 21, no. 3, pp. 589–594, 2006.

[JOH 04] JOHNSON G.R., FERRARIN M., HARRINGTON M., *et al.*, "Performance specification for lower limb orthotic devices", *Clinical Biomechanics*, vol. 19, no. 7, pp. 711–718, 2004.

[KAR 11] KARK L., SIMMONS A., "Patient satisfaction following lower-limb amputation: the role of gait deviation", *Prosthetics and Orthotics International*, vol. 35, no. 2, pp. 225–33, 2011.

[LAM 05] LAM W.K., LEONG J.C.Y., LI Y.H., *et al.*, "Biomechanical and electromyographic evaluation of ankle foot orthosis and dynamic ankle foot orthosis in spastic cerebral palsy", *Gait & Posture*, vol. 22, no. 3, pp. 189–197, 2005.

[LOR 07] LORENA A.C., CARVALHO A.C.P.L.F., "Protein cellular localization prediction with support vector machines and decision trees", *Computer in Biology and Medicine*, vol. 37, pp. 115–125, 2007.

[MAV 11] MAVROIDIS C., RANKY R.G., SIVAK M.L., *et al.*, "Patient specific ankle-foot orthoses using rapid prototyping", *Journal of NeuroEngineering and Rehabilitation*, vol. 8, pp. 1–11, 2011.

[MCG 09] MCGINLEY J.L., BAKER R., WOLFE R., *et al.*, "The reliability of three-dimensional kinematic gait measurements: a systematic review", *Gait & Posture*, vol. 29, no. 3, pp. 360–369, 2009.

[ROS 03] ROSSE C., MEJINO J., "A reference ontology for biomedical informatics: the foundational model of anatomy", *Journal of Biomedical Informatics*, vol. 36, pp. 478–500, 2003.

[REC 91] RECTOR A., NOLAN W.A., KAY S., "Foundations for an electronic medical record", *Methods of Information in Medicine*, vol. 30, pp. 179–186, 1991.

[RUB 07] RUBIN D.L., NOY N.F., MUSEN M.A., "Protégé: a tool for managing and using terminology in radiology applications", *Journal of Digital Imaging*, vol. 20, no. 1, pp. 34–46, 2007.

[SCH 03] SCHNIERDER L., "Foundational ontologies and the realist bias", *KI2003 Workshop on Reference Ontologies and Application Ontologies*, Hamburg, Germany, 2003.

[SIL 11] SILVER-THORN B., HERRMANN A., CURRENT T., *et al.*, "Effect of ankle orientation on heel loading and knee stability for post-stroke individuals wearing ankle-foot orthoses", *Prosthetics and Orthotics International*, vol. 35, no. 2, pp. 150–162, 2011.

[SMI 07] SMITH B., ASHBURNER M., ROSSE C., *et al.*, "The OBO foundry: coordinated evolution of ontologies to support biomedical data integration", *Nature Biotechnology*, vol. 25, no. 11, pp. 1251–1255, 2007.

[STA 01] STAAB S., SCHUUR H., STUDER R., *et al.*, "Knowledge processes and ontologies", *IEEE Intelligent Systems*, vol. 16, no. 1, pp. 26–34, 2001.

[STE 07] STEFAN S., UDO H., "Towards the ontological foundations of symbolic biologicalTheories", *Artificial Intelligence in Medicine*, vol. 39, pp. 237–250, 2007.

[THO 07] THOMAS B., MAUREEN D., "Logical properties of foundational relations in bio-ontologies", *Artificial Intelligence in Medicine*, vol. 39, pp. 197–216, 2007.

[TUR 13] TURCIN I., ERGOVIC V., LACKOVIC M., "Ontology driven decision support system architecture for gait analysis", *IFMBE Proceedings*, vol. 38, pp. 78–81, 2013.

[WAK 07] WAKELING J., DELANEY R., DUDKIEWICZ I., "A method for quantifying dynamic muscle dysfunction in children and young adults with cerebral palsy", *Gait & Posture*, vol. 25, pp. 580–589, 2007.

[ZWE 95] ZWEIGENBAUM P., BACHIMONT B., BOUAUD J., *et al.*, "Issues in the structuring and acquisition of an ontology for medical language understanding", *Methods in Informatics Medicine*, vol. 34, nos. 1–2, pp. 15–24, 1995.

5

Software and Tools for Knowledge Modeling and Reasoning/Inference

This chapter presents the software and computing tools that could be used for practical works regarding the knowledge modeling and reasoning/inference functionalities of a specific application. A non-exhaustive list of open source and commercial software and tools will be presented. Two open source tools (Protégé and Java Expert System Shell (JESS)) will be focused on and described in more detail for practical purpose.

5.1. Open source and commercial knowledge modeling software and tools

5.1.1. *Open source*

Open source is one of the most important advancements in the development of the software community from the 1960s. Open source aims to provide universal access to a specific software application or programming library for a wide range of users throughout the world. A computer program under the open source label provides the source code to all users for possible use and/or eventual modification from its original design. The concept of open source is always linked to the collaborative effort to create, update and share the code with the community of interest.

A computer program under the open source label has a copyleft free software license that gives the users all freedoms in downloading, use, modification and redistribution of the software. GNU General Public License (GPL) is an example of copyleft free software license.

5.1.2. *List of open source software and tools for knowledge modeling*

Knowledge modeling and management plays an essential role in the formalization and sharing of domain-based knowledge in a specific field of study. Thus, the use of open source software and tools accelerates the impact of the knowledge modeling and management within the community. A non-exhaustive list of the most widely used open source software and tools is provided as follows:

– *Protégé* is a free, open source ontology editor and knowledge-based framework [STA 13].

– *Freeplane* aims to develop, organize and communicate the ideas and knowledge [POL 13].

– *Knowledge Mapper* is an open source tool for document repository, knowledge extraction or categorization [SOU 13a].

– *TextToOnto* aims to apply text mining techniques in the ontology construction process [SOU 13b].

– *Exteca* is a software solution for developing knowledge models to support knowledge management systems [FER 13].

– *JESS with academic license* is a rule engine and scripting environment [SAN 13].

– *Apache Jena* is a free and open source for building semantic web and linked data applications [APA 13].

5.1.3. *List of commercial software and tools for knowledge modeling*

Open source software and tools could be downloaded and used free of charge but the main disadvantages are the non-supportive character and heterogeneous updates of new components due to the lack of well-controlled processes. Commercial software and tools allows a continuous support and well-controlled update process for the end users. A non-exhaustive list of the most widely used commercial software and tools is provided as follows:

– *JESS with commercial license* is a rule engine and scripting environment [SAN 13].

– *SAS® Text Miner* aims to discover information and knowledge such as themes, patterns or emerging issues in document collections [SAS 13a].

– *SAS® Ontology Management* is a technical solution to define and manage semantic terms [SAS 13b].

– *WordMapper* is a text mining solution [FBC 13].

– *IBM InfoSphere Data Explorer* is an information and content discovery and management tool [IBM 13].

– *Be Informed Suite* is a tool for building large ontology-based applications (visual editors, inference engines, standard formats exportation, etc.) [BE 13].

5.2. Protégé: ontology editor and knowledge-based framework

5.2.1. *Introduction*

Among the most widely used knowledge modeling tools for biomedical applications is the Protégé software. Protégé is a Java-based ontology editor developed at Stanford University in collaboration with the University of Manchester. This software is made available under the Mozilla Public

License 1.1. Protégé is a powerful and widely used tool for knowledge creation and management as quoted in the official Website (http://protege.stanford.edu/): *"Protégé is supported by a strong community of developers and academic, government and corporate users, who are using Protégé for knowledge solutions in areas as diverse as biomedicine, intelligence gathering, and corporate modeling"*. The interaction between the Protégé platform and the end user to model a specific ontology is managed via a web client or a desktop client interface. There are three ontology formats supported by the Protégé platform such as OWL, RDF(S) and XML schema. Moreover, the Protégé platform is extensible, thus the end user could include additional components into the Protégé platform. The current version of the Protégé platform is 4.3 (November 2013). The main interface of the Protégé platform is illustrated in Figure 5.1.

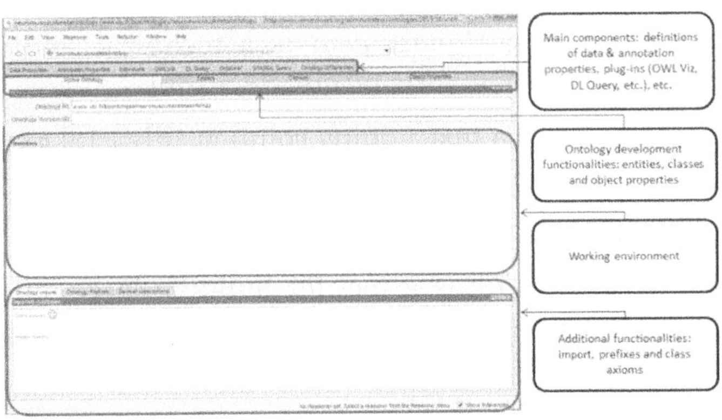

Figure 5.1. *Protégé platform: main interface of version 4.3*

5.2.2. *Ontology development methodology*

The development of an ontology using the Protégé platform consists of the following steps [NOY 13]:

– Step 1: determine the domain and scope of the ontology;

– Step 2: consider reusing existing ontologies;

– Step 3: enumerate important terms in the ontology;

– Step 4: define the classes and the class hierarchy (Figure 5.2);

– Step 5: define the properties of classes slots;

– Step 6: define the facets of the slots;

– Step 7: create instances.

Figure 5.2. *Class definition within Protégé platform*

5.2.3. *Bio-ontology example*

To show the different functionalities of the Protégé platform in the creation of a bio-ontology, a simplified neuromusculoskeletal ontology was designed and created. The neuromusculoskeletal ontology consists of two main classes: the nervous system and musculoskeletal system. The musculoskeletal system class consists of two subclasses: muscle and bone. The nervous system consists of one class: motor unit (as shown in Figure 5.3). An "attach" object property was created between muscle and bone classes to define their anatomical relationship (as shown in Figure 5.4). The morphological and mechanical properties of muscles were created and linked to muscle class using the data property as shown in Figure 5.5.

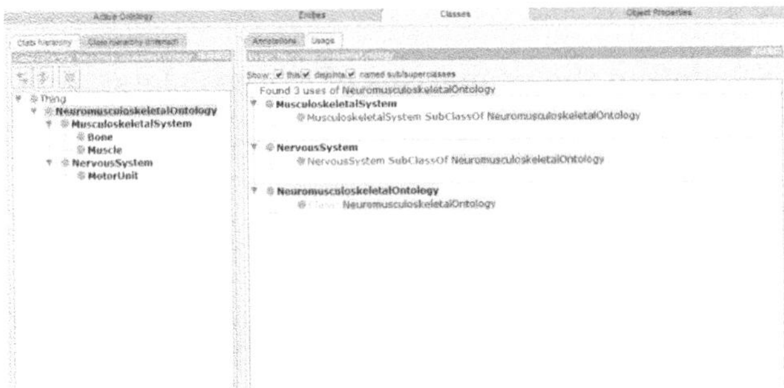

Figure 5.3. *Classes of the neuromusculoskeletal ontology*

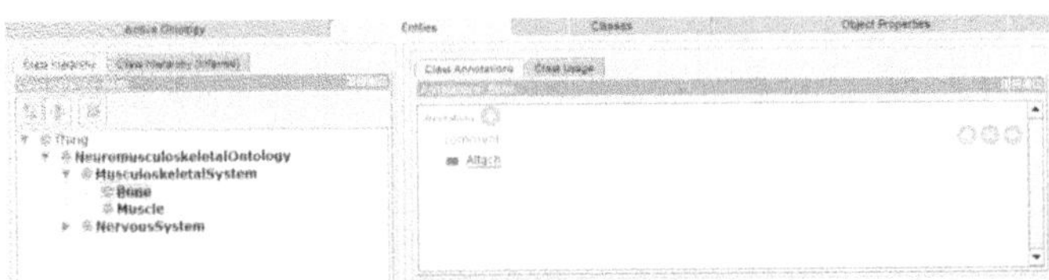

Figure 5.4. *"Attach" object property to define the anatomical relationship between bone and muscle classes*

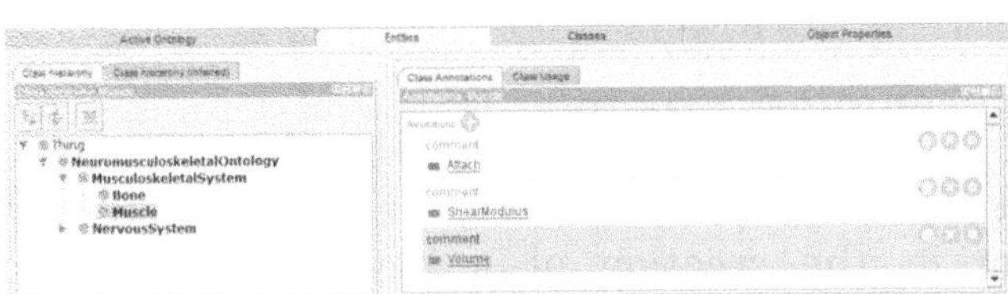

Figure 5.5. *Data properties of the muscle class*

5.3. JESS: reasoning and inference library

5.3.1. *Introduction*

JESS is a Java-based programming library developed by Ernest Friedman-Hill at Sandia National Laboratories. This library is a rule engine dedicated to Java-based reasoning/inference applications. Using JESS, new facts/knowledge could be deduced from existing facts/knowledge in the form of declarative rules. JESS engine implemented a well-known Rete algorithm [FOR 82] to fire the new rules from the fact database.

5.3.2. *Development process of a rule engine*

The development of a rule engine program needs the following steps:

– Step 1: define all initial facts as declarative and production rules;

– Step 2: develop a reasoning motor;

– Step 3: enter facts to fire the rule set;

– Step 4: extract new facts/knowledge from firing process.

5.3.3. *Example*

The example presented here is a simplified medical diagnosis of muscle spasticity using morphological and mechanical properties. It is assumed that current knowledge about muscle spasticity is presented in Table 5.1.

Facts	Rules
Muscle has a thickness	R1: If muscle thickness < 5 cm^2, then Fire R2
Muscle has a volume	R2: If muscle volume < 10 cm^3, then Fire R3
Muscle has a shear modulus in passive or active conditions	R3: If muscle passive shear modulus < 1 kPa, then muscle spasticity = severe

Table 5.1. *Example of facts set and rules database*

The diagnosis process is based on the values of morphological and mechanical properties of the patient. An example of values and fired rules are presented in Table 5.2.

Working facts of patient X	Fired rules
Muscle thickness = 4 cm^2 Muscle volume = 8 cm3 Muscle passive shear modulus = 0.8 kPa	Fired R1, R2 and R3 Result: Muscle spasticity = severe

Table 5.2. *Example of fired rules using facts*

5.4. Conclusion

This chapter presents some open source useful software and tools to frequently extract, create, formalize and reason about the knowledge of a domain of interest. Two main tools (Protégé and JESS) were described in detail with illustrative examples.

5.5. Summary

– Open source software and tools provide universal access to a computer program.

– An open source computer program could be downloaded and used free-of-charge. It could be updated, modified and redistributed by any end user.

– Protégé is a Java-based ontology editor platform.

– JESS is a Java-based rule engine allowing backward and forward chaining to be executed to deduce new facts/knowledge about a domain of interest.

5.6. Bibliography

[APA 13] APACHE SOFTWARE FOUNDATION, 2013. Available at http://jena.apache.org/.

[BE 13] BE INFORMED USA, INC., 2013. Available at http://www.beinformed.com.

[FBC 13] FBC SOFTWARE, 2013. Available at http://www.fbcsoftware.com/index.php/wordmapper-entreprises.html.

[FER 13] FERNANDES L., TALEVI M., JAIN N., *et al.*, 2013. Available at http://exteca.sourceforge.net/.

[FOR 82] FORGY C.L., "Rete: a fast algorithm for the many pattern/many object pattern match problem", *Artificial Intelligence*, vol. 19, pp. 17–37, 1982.

[IBM 13] IBM, 2013. Available at http://www-03.ibm.com/software/products/en/dataexplorer/.

[NOY 13] NOY N.F., MCGUINNESS D.L., "Ontology development 101: a guide to creating your first ontology", 2013. Available at http://protege.stanford.edu/publications/ontology_development/o ontology101.pdf.

[POL 13] POLIVAEV D., 2013. Available at http://sourceforge.net/projects/freeplane/.

[SAN 13] SANDIA NATIONAL LABORATORIES, 2013. Available at http://herzberg.ca.sandia.gov/.

[SAS 13a] SAS INSTITUTE INC., 2013. Available at http://www.sas.com/text-analytics/text-miner/.

[SAS 13b] SAS INSTITUTE INC., 2013. Available at http://www.sas.com/text-analytics/ontology-management/.

[SOU 13a] SOURCE FORGE COMMUNITY, 2013. Available at http://kmapper.sourceforge.net/.

[SOU 13b] SOURCE FORGE COMMUNITY, 2013. Available at http://texttoonto.sourceforge.net/.

[STA 13] STANFORD CENTER FOR BIOMEDICAL INFORMATICS RESEARCH, 2013. Available at http://protege.stanford.edu/.

Index

S, T, U, V, X

Printed and bound by CPI Group (UK) Ltd, Croydon, CR0 4YY

13/09/2026

14961082-0005